Quantum Theory

Quantum Theory

A Philosopher's Overview

Salvator Cannavo

Published by State University of New York Press, Albany

Printed in the United States of America

For information, contact State University of New York Press, Albany, NY
www.sunypress.edu

Production by Kelli W. LeRoux
Marketing by Anne M. Valentine

Library of Congress Cataloging-in-Publication Data

Cannavo, S. (Salvator)
Quantum theory : a philosopher's overview / Salvator Cannavo.
p. cm.
Includes bibliographical references and index.
ISBN 978-0-7914-9347-2 (hardcover : alk. paper)
1. Quantum theory—Philosophy. I. Title.

QC174.13.C36 2009
530.12—dc22 2008027670

10 9 8 7 6 5 4 3 2 1

Contents

Preface

My interest in the foundations and philosophy of quantum theory was first sparked by J. M. Jauch who taught our graduate course in quantum mechanics at Princeton University. But this was only the beginning. The seminars and teas that were occasionally held in Fine Hall and which we graduate students were privileged to attend were the ultimate in stimulation and fascination. The participants were Princeton faculty and members of the Institute for Advanced Studies. And it was an unforgettable experience to see some of the very architects of quantum theory express themselves impromptu on how to understand or "interpret" it. Most remarkably the issues that arose then are the issues of interpretation that persist to this very day. In essence, therefore, the conceptualization of this book began at that time.

The actual writing, however, jumpstarted with my own lectures on quantum theory at the City University of New York. My thoughts in this area evolved further with papers in the philosophy of science that I read to the Teilhard de Chardin Conference at Brooklyn College of CCNY, to the Royal Institute of Philosophy at the University of Southern California, to the Long Island Philosophical Society, and to the Academy for the Humanities and the Sciences at the Graduate Center of the City University of New York. This last lecture was a lead paper on "The Reality Problem in Quantum Mechanics," and it was this reading followed by the long and searching philosophical discussion that followed that helped shape the main themes in this book.

Among the features that any scientific theory about the physical world is expected to have, two are basic, though not both required. First, the theory must make *prediction* possible. That is, by means of its mathematical formalism and, on the basis of the initial state of a physical system at some given time, the theory must enable one to compute the state of that system at some future time. Secondly, but not required, a theory *may* also be expected to provide for *explanation.*

That is, it may be expected to enable one to say, *either in causal terms* or in terms of some sort of *unification,* how the change from one state to another comes about. While all physical theories must predict successfully (this goes into granting them confirmation and eventual acceptance), not all can (nor must they) explain what they predict.

In this general regard, one need hardly note that the notion of "explaining" ranges over a variety of senses, such as that of subsuming under more general principles, explicating, justifying, and more. This, however, need not delay us. Explaining, either in causal terms or in terms of some sort of unification is *one* of these senses and the one concerning us here.

In order for a physical theory to explain. it must, as an addition to its formalism, make the existential or, as we might say, "ontological" claim that there exist physical entities to which some terms of the theory make general reference. The entities must be either capable of interacting dynamically over space and time (e.g., masses, particles, forces, charges, fields) or they must be relatable in other essential respects. Moreover, in order that the resulting explanations do what they're supposed to do—namely, enlighten our grasp or understanding of the "how and why" of things—the posited existential subject matter must itself be coherent and therefore intelligible. This means it must, at least in some respects, be observer independent or objective, existentially stable, uniquely locatable in physical space-time, and *preferably* susceptible only to proximate or local influences rather than to unmediated, instantaneous, and therefore, mysterious ones. To this, one might cautiously add that what an *explanatory* theory tells us about the physical world must, to some sufficient extent, be intuitable. Indeed, what we are describing here is what one might more succinctly call a "reality" model, and the coordination of such a model with the formalism of a theory is what we are calling the *ontology* or *interpretation* of the theory.

After one hundred years of astoundingly successful prediction, quantum theory has yet to find itself such an interpretation in this sense and on which there is general agreement. Paradoxically, therefore, quantum theory—perhaps the most elegantly structured mathematical edifice in the history of science—predicts but does not *explain* what it so accurately predicts.

Some might want to say at this point that the idea of framing a physical theory in such a way as to invest it with the sort of content we are describing and therefore with explanatory power is no more than the arbitrary imposition of classical notions and content on an essentially nonclassical theory. We shall argue, however, that any such concern would be a misguided one. The power of a theory to explain is not merely a feature limited to classical physics operating in classical contexts but a feature for delivering something essential to an understanding of the physical world—something epistemically basic and quite apart from any distinction between classical and nonclassical modes of description. In fact there are, as will be discussed in the main text, quantum-type theories, for example, quantum electrodynamics, quantum field theory, and string theory that—whatever their present stage of development—are *explanatory* in a sense that is philosophically satisfying, though requiring a degree of liberalization or "tolerance" in the underlying realism.

Though falling short of sufficient physical content for explanation, quantum theory must nevertheless be tied to observational subject matter (scintillations, clicks, photographic patterns, etc.). If it weren't, it would be not physics but only a mathematical system for relating abstract symbols. This subject matter, however, does not, by itself, comprise a coherent physical model or interpretation. True, quantum-theoretic terms, are customarily understood to refer to a variety of entities and their attributes (particles, spin, momentum, location, energy, etc.). This "quantum subject matter," however, immediately tests our understanding with its strangeness. We find ourselves dealing with such physical oddities as entities that have no unique locations, particles that have spins but no dimensions, unsettling discontinuities and much, much more to perplex us. Indeed, the quantum formalism, which, in itself, seriously blocks *physical* interpretations, is framed in a space of variable dimensions whose abstract constructions cannot be simply coordinated with much more than raw observational results and the parameters of given experiments. Any attempt, for example, to assign concrete existential status to the so-called "matter waves" sometimes associated with the very heart of quantum theory seems hopeless. The dynamical features required of vibrating physical systems simply do not apply to these abstract "waves." And all other versions of the quantum mechanical formalism are similarly unreceptive to dynamical interpretation.

The difficulties deepen as theorists try to make better sense of the reality that underlies the phenomena that the quantum formalism enables us to predict. What they come up with is a microworld that is causally unruly, situationally and existentially elusive, and clashes head-on both with our most basic intuition as well as with the objective realism that rules not only common sense but, demonstrably, the deepest philosophical attitudes of virtually all natural scientists. Indeed, quantum predictions and some of the basic theorems of the quantum formalism such as, for example, the principles of indeterminacy and superposition, run not only counter to highly entrenched classical principles, but—more seriously—also pose unsettling issues about the very intelligibility and objectivity of all human experience. The great enigma at the foundations of quantum theory, therefore, has been (as Hillary Putnam once pointed out in somewhat different words): *Why should an ultra-abstract formalism whose predictions lead to such perplexing and hugely counterintuitive visions of reality be so successful?*

This gnawing question, which foundational theorists refer to as *the problem of interpretation,* has fueled a century-long history of extensive controversies that have yielded no generally accepted result. Two interpretations, the *Copenhagen* and *hidden variable* interpretations, have been the most prominent attempts at a solution, but they have spawned a swarm of difficulties of their own. More recent interpretations such as the *many worlds* and *many minds* interpretations are so highly imaginative and belief-straining as to have generated more polemics than acceptance, to the point of even drawing the charge of grotesqueness from some of the most positivistically inclined critics.

This book is an overview that attempts a fresh perspective on the long-standing problem of making "explanatory sense" of quantum theory or, in more standard terms, *the problem of interpretation.* In the course of examining the record on this highly resistive problem, the discussion encounters the most pressing, and as yet unsettled concerns in the foundations of quantum theory. It outlines the failing attempts at resolving the main issues, highlights some of the underlying philosophical puzzles, and finally arrives at a proposal for the direction of future theorizing—a proposal inspired by modern explanatory theories such as quantum electrodynamics, quantum field theory, and string theory.

Chapter 1 starts with the well-known experimental and conceptual circumstances that sparked quantum theory. It then goes on to trace the rapid and odd genesis that leads to *quantum mechanics*—an elegant ultra-abstract formalism uniquely devoid of "solid" explanatory anchorage, yet astoundingly successful as a mathematical engine for prediction.

The second and third chapters review the major puzzles that have dogged students of quantum theory since the very beginning of its history. The second chapter first endeavors to take the reader through a nontechnical examination of the basic or "core" formalism of quantum mechanics. The third chapter then goes on to examine key aspects of the theory as a whole, together with the problems they raise. Such baffling quantum features as wave-collapse, superposition, phase entanglement, virtual particle creation, and the tradition-shattering possibility of "remote" or *non-local* causation are as philosophically intriguing as they are mindbending.

As we get to the fourth chapter, we are ready for a critical look at the major interpretations of quantum theory. These are *the hidden variable* and *the Copenhagen interpretations.* Both come early in the history of the subject, and neither has met with general acceptance, though the Copenhagen interpretation is dominant among contemporary physicists. The account aims to bring out serious failings in both interpretations. For example, it indicates how the Copenhagen interpretation flatly deprives quantum theory of the "full" and *objective* kind of ontology necessary for explaining what is "really" happening beneath the experimental "appearances." In this chapter, we also introduce what is perhaps the deepest and most intractable puzzle at the foundations of quantum theory, namely, *the measurement problem.*

The discussion then goes on to show how the Copenhagen interpretation leaves the physical world sundered in two by failing on this problem. Von Neumann's attempt to fix all this with his add-on notion of collapsing waves is also shown to be itself too troublesome (physically and, one might say, metaphysically) to settle the issue. The second of the two interpretations under consideration, namely, the *hidden variable interpretation,* is also seen to fail. It not only raises serious questions regarding the physical significance of its unobservable or "hidden" variables, but it runs into logically gnawing issues of its very possibility.

In the wake of this account, chapter 5, titled,"Fresh Starts," turns to more recent interpretations, namely, the *many worlds interpretation* (MWI) and the *many minds interpretation* (MMI), *decoherence theory* (DT), and *GRW theory.* These, we maintain, also fare badly. While they are highly imaginative and even fascinating, they are metaphysically burdened to the point of raising serious questions about testability and physical significance. Moreover, they are so convoluted and stray so wildly from any sort of common sense that they test the very limits of belief and have even drawn the charge of being extravagant. What is perhaps a more serious philosophical objection is the fact that these interpretations—especially MMI—tend to make gratuitous metaphysical assumptions about the mind-body relation without the benefit of any established theory of consciousness. DT and GRW theory are more promising but the discussion attempts to show that they too fall short of the desired goal.

Chapter 6 begins with a key distinction between two kinds of scientific laws and theories: those that not only predict but also *explain* in terms of entities such as particles, forces, fields, etc. and those that, lacking such dynamical subject matter, predict without explaining. We refer to laws and theories that explain as *explanatory nomologicals* and those that predict without explaining as *algorithmic nomologicals.*

In the context of this distinction, chapter 6 finds quantum theory to be *ultimately* an algorithmic device for prediction. Indeed, except for some early beginnings, quantum theory is shown to have been born and subsequently fashioned with virtually no concerns for positing any sort of dynamical (i.e., causally functioning) subject matter on the basis of which to make "physical, explanatory sense" of quantum phenomena. "Probability waves" which superficially resemble dynamical processes are *not* concrete physical phenomena, and even the quantity "electron spin"—a quantity derivable from Dirac's new formulation of the theory in 1928—is only as a heuristic analogy. Actual rotational spin cannot be consistently predicated of the dimensionless electron. The electron therefore behaves only *as if* it had a physically possible or "actual" angular momentum and an associated magnetic moment.

At this point in tracing the development of the quantum formalism, chapter 6 gives some passing attention to Dirac transformation theory not only to note its remarkable power and applicability but also its foreshadowing of quantum electrodynamics, a genuinely explanatory theory that the great quantum physicist, M. Dirac, launched. The chapter ends with some discussion of the very strong role of symmetry in scientific theorizing but observes that the appeals to symmetry in the development of quantum theory are too abstract to provide explanatory power.

The starkly and irreducibly algorithmic nature of quantum theory and the very long history of failed efforts at formulating a generally acceptable interpretation strongly suggest that the interpretational program is deeply impractical, if not utterly futile—at least, for quantum theory in its present form. Accordingly, chapter 7 raises the dire question of whether to abandon this failure-prone program altogether and leave the explanatory physics of nature to new theories equipped with the appropriate existential (ontological) content. Would such a move entail jettisoning quantum theory, or can quantum mechanics be conjoined as an algorithmic *ancillary* context to enhance the predictive power of some new explanatory theory? What can we learn in this regard from *quantum electrodynamics, quantum field theory,* and *string theory* all three of which are monumental examples of such a fusion?

With these questions on the table, chapter 8 undertakes a summative nontechnical examination of these three theories. The discussion, however, adds a cautious reminder. There is a price for the quantum context that these theories appropriate. The dramatic developments of several decades ago, relating to a discovery by John Bell, require a shift in the standard view of causation: *A physics that accepts quantum theory—even as only the context for a new theory—must also accept the possibility of non-local influences,* that is, the possibility of what Einstein regarded as "spooky" action at a distance. And this means some philosophical departure from the traditional commonsense realism of natural science.

The closing sections of the book, namely, chapter 9, give a passing consideration to one remaining alternative for resolving the foundational issues of quantum theory. This would be the replacement of quantum mechanics by a new integrative theory of the physical world. Historically, this has been both recommended and tried, and the discussion offers some general comments on these efforts and on future possibilities.

The discussion in the main body of text is primarily philosophical and the mode of exposition aims to be ordinary language or, as one might say, informal and nontechnical. For lucidity and accessibility, I have tried, throughout, to say how I was using the very few special terms that unavoidably found their way into the main text. For more completeness and precision, however, I thought it helpful to include a few technical and mathematical elaborations. These, however, are virtually all confined to the endnotes so that the reader may readily omit them without losing continuity or the line of argument.

The ordinary language mode of exposition should make the main body of discussion accessible to the general reader with a mature interest in science while the elaborations in the endnotes may engage those who are more mathematically inclined. As a historical and philosophical overview, the text may be found suitable for courses in the history and philosophy of science, and, as a reading for university-level seminars in quantum physics.

The encouragement and inspiration for this book have come from many sources.

First I must express the deepest of my indebtedness to the titanic masters of the subject, Einstein, von Neumann, and Pauli, who at the time of my graduate years were at the Institute for Advanced Studies at Princeton. I had the privilege of attending their seminars, and their comments on the very subject to which they had so vastly contributed were a source of enlightenment and inspiration. At that same time I also profited from my frequent discussions with Richard Feynman and with distinguished fellow students such as O. Halpern, J. Keller, and R. Bellman—all later contributors to the subject. On the *problem of quantum measurement, I* profited especially from my association with H. Everett, author of the many worlds interpretation. It was he that first made me aware of the problem when we were both Ford Foundation Fellows at Wesleyan University.

These were all physicists but I owe as much to philosophers, some of whom were my teachers and others my colleagues in the New York circle of philosophers. Among these I must mention Professors E. Nagel, C. Hempel, S. Morgenbesser, S. Hook, A. Hofstadter, and visiting professor A. J. Ayer, each of whom on several occasions of shared discussion with me contributed much to what I was doing in one of the then central subjects of the philosophy of science, namely, scientific explanation.

I must also mention my most recent benefactors. First I wish to express deep gratitude to my colleagues, Professors Michael Sobel, K. D. Irani, Daniel Greenberger, and Wolston Brown for patiently reading through the entire manuscript and for their detailed and invaluable comments. In this regard, I cannot fail to mention my peer reviewers—of unknown identity—whose comments also helped finish the final draft. The errors or failures of insight that may still remain, however, are, of course, entirely mine.

At SUNY Press, I would like to thank Jane Bunker, editor-in-chief and her assistant, Andrew Kenyon, for their patient editing in the face of very challenging detail and unwavering faith in my project, as they guided it through the publication process. Also, I cannot fail to thank Kelli LeRoux, senior production editor and Anne Valentine, executive promotions manager for their kind helpfulness in dealing with all my many late changes.

Precious encouragement and ideas throughout the years of writing and research were always near at hand in my own family. For these I must thank my children, Dr. Francesca Cannavo Antognini, Dr. Joseph P. Cannavo, and Professor Peter F. Cannavo. Finally, my deepest debt of love and gratitude goes to my dear wife, Gaetana, poet and educator, whose editorial acumen helped set the language of the text. Indeed, without Gaetana's loving support this manuscript would not now be.

Before closing, I must also say something about the excitement of the project. Diderot once said, "If you want to learn a subject, teach it." He might have added, "or write a book on it." The philosophy of quantum theory can, in places, pose hard technical challenges. But navigating some of the tracts, which a century of "quantum polemics" has generated, has been an edifying and mind-clearing experience. Even more than this, however, it has been an exciting, though daunting, adventure.

1

The Quantum and Classical Theories

A Crucial Difference of Theory-Type

Quantum Theory

Quantum theory (QT) has been with us for more than one hundred years, its earliest beginnings going back to the start of the twentieth century. It was then that Max Planck, after much effort to make theory fit fact, published his quantum hypothesis in order to provide some sort of basis for all that was then known experimentally about black body radiation.[1] Einstein's light corpuscle followed four years later to explain the photoelectric effect, and hardly two decades later, a mathematical quantum framework emerged whose predictive success was nothing short of spectacular.

Almost immediately, *quantum mechanics*—as this framework came to be known—yielded predictions of the known facts of spectroscopy, the Zeeman, Stark, and Compton effects, scattering phenomena, photo electricity, and the periodic table. The new "mechanics" promised to fill gaps in a classical theory that could not account for the specific heats of solids and the stability of the Rutherford planetary atom. Several decades later, theoreticians developed *quantum electrodynamics*—an electro-dynamical theory whose quantum-theoretic concepts of spin and resonance shed light on magnetism, chemical bonds, and more. Before long, QT was a scientific sensation as its predictions became more and more testable to a very high level of precision with ever more sophisticated experimental equipment.

Presently, QT is fully operative in solid state physics and therefore at the cutting edge of some of our cherished transistor, semiconductor, superconductor, and computer hardware technologies. It has found applications in laser, cryogenic, and genetic engineering as well as in theories of electrical conductivity, magnetic materials, and changes of state. More recently, it is providing the developmental basis for new technologies in quantum computing and cryptography involving some of the most arcane aspects of microphenomena. Only a few of the many possible examples of such applications are: *tunneling* for tunnel diodes, *superposition* at the level of atomic spin in order to vastly increase the number of simultaneous computing tasks, *phase entanglement* for correcting computer errors, and *decoherence* as a mode of control in quantum computing. There is also the application of *electron pairing* in superconductivity, in Bose-Einstein condensates, and in neutron beam thermometers. Some experiments even suggest the possibility of future micromachines using the ultrasmall Casimir force due to the pressure of virtual photons. And though cosmology has not yet systematically integrated with QT, the supplementation of general relativity theory with some quantum-theoretic rules vastly extends our reach into the elusive nature of black holes along with the possible origins and future of the universe. QT has thus figured crucially in modern physical theory from the microlevel of beta decay to the cosmic one of star formation.

But even more than all this, QT provides a contextual framework for quantum electrodynamics and quantum field theory now capable of encompassing the three nongravitational forces (weak, strong, and electrical) so basic to our understanding the causal dynamics of the physical world. Also noteworthy are the more recent efforts to bring the fourth basic force, gravity, into a final unification. These have resulted in various forms of *string theory*, still in unfinished states, but again developed as "quantum-type" or, as we shall call them, *quantumized* theories.[2] Finally, out of attempts to unify these forms, has emerged M-theory, presently only gestating, but with the promise of a vastly unprecedented explanatory scope. So, by all indications, QT is here to stay—at least for a long, long while. Indeed, its sweeping success in predicting microphenomena, together with its striking mathematical elegance, have earned it a top place among the most monumental of scientific creations.

Classical Theory

The landmark difficulties that beset classical theory almost a century ago—in special areas such as radioactivity, photoelectricity, black body radiation, specific heats, atomic spectra, and atomic structure—were certainly real and compelling signs that fresh approaches to the physics of nature were critically needed. Despite this crisis, however, classical theory—meaning by this, all of physical theory other than QT and *quantum-type (quantumized)* theories—remained (and continues to be) immensely successful in dealing with a vast range of phenomena including virtually all that happens in the familiar world of everyday life.

We have, here, a level of success that has, time and again, been nothing short of spectacular. In this regard, some all too familiar examples still bear mention, namely, the prediction of the existence of Neptune and of its orbit in the mid-nineteenth century (before astronomers physically discovered the planet), or the spectacle of wireless communication at the start of the twentieth century—all on the basis of classical principles. Indeed, quite apart from the thunderous technological impact of quantum physics in some basic areas, the vastly major portion of present day macrotechnology from computing to space science, continues to be classically based. Also to be noted, in this regard, are the dramatic successes of relativity in both applied and theoretical contexts. They have been awesome, especially given some of the intuition-straining content and predictions of the theory. (From the standpoint of the present discussion, we regard relativity theory as essentially classical.)

The "Unfinished" Quantum Theory

None of this glorious classical history, however, has been quite as stunning as the success of QT—a success that has seemed utterly magical. The reason for the uncanniness, paradoxically and curiously enough, is what some see as an "unfinished" state of the theory. When compared to most great scientific theories, QT is missing something—something traditionally deemed important for any theory of nature to have in order to explain the facts of experience. What QT has not yet found itself is a universally accepted interpretation, and this means that there is no generally settled opinion on the kind of existential subject matter or, if you like, ontology to commit to, in order to provide the theory with explanatory power.

On reflection, however, it need not be so surprising that finding an interpretation of QT, especially one structured for explanation, is problematic. To begin with, the nature of scientific explanation is a controversial issue in the philosophy of science with disagreements reflecting widely differing philosophical orientations, for example, positivist, realist, etc. But more than this, the formalism of QT is itself, in some respects, remarkably resistant to physical interpretation. QT is irreducibly statistical. It is not about what will actually happen to any concrete physical system. It is about *possible* states (quantum states) in which not all the constitutive variables have determinate values on the basis of which to explain, or even just predict, any definite happening, that is, one involving individual physical entities. More specifically, what the theory defines and predicts is no more than the probabilities of such *possible* quantum states—probabilities grounded on expected relative frequencies in the outcomes of measurement.

Adding to this bleakness is the somewhat disconcerting fact that the results of a long history of investigations seem to block, in principle, any possibility of remediation by supplementing QT with additional subject matter, namely, additional variables. These would be variables on the basis of which (1) to give determinate and intelligible physical accounts of what is happening behind the statistical appearances, and (2) to recover, by averaging methods, the statistical predictions of QT. It has been variously shown on the basis of what are known as "no-go theorems" that such a "hidden variable approach"—as it is called—is not possible without violating both a reasonable measure of basic realism and some rock-bottom requirements of commonsense intelligibility. That is, no such interpretation of QT could possibly satisfy certain restraints required by any realism of the kind insisted upon by Einstein throughout a good part of his lifetime and, one may add, also required by the standards of ordinary commonsense intelligibility.[3]

Still other obstacles hamper agreement on any interpretation. The statistical predictions of QT regarding the data we get in quantum experiments are indeed remarkably accurate and refined. The moment we try to explain these results in terms of some underlying reality, however, we come up with bewildering scenarios. These are so vastly counterintuitive and so violating of common sense that even the most distinguished contributors are unable to come to terms on any universally acceptable existential framework for "grasping" what is going on behind the experimental appearances. Add to this the further hobbling of any agreement not only by differing philosophical attitudes but also by the logical impasses we have mentioned, and the issue of interpretation seems to become virtually irresolvable.

The question that remains therefore is: How do we define (describe, characterize) quantum subject matter? Obviously, the answer that will inform our grasp must be one framed in minimally understandable terms, that is, in terms of the familiar concepts that *consistently* structure what ordinarily counts as "real" and intelligible. Things don't simply go in and out of existence; they change in causal contexts, and they never present simultaneously incompatible traits.

More generally, what we have in mind here are such "reality features" as substance stability or conservation, continuity of change, causality, and objectivity or mind-independence. To this we might add a measure of observer independence on the basis of the doctrine that one can, at least in principle, correct for any disturbance of the observed subject matter by the act of measurement. These, then, are the sort of "reality attributes," in terms of which one might want to frame any account of quantum subject matter. The idea is that such an account would be encompassing and coherent enough for making sense of the often astounding predictions of QT—predictions that experiment so remarkably confirms. Indeed, absent anything like such a framework or interpretation, the predictions of the quantum formalism, however accurate, remain stubbornly and opaquely puzzling.

As we intend it here, our notion that an interpretation of QT, in order to make sense of its predictions, must incorporate intelligibly structured physical subject matter, is not a bald claim about any "ultimate metaphysical status" of physical reality (mentalistic, materialistic, or other). Rather, it is a purely epistemic requirement about the sort of properties and relational attributes in terms of which we can say that we understand QT. Similarly, our subject matter requirement is independent of any distinction some might wish to draw between so-called quantum and classical modes of description. Our requirement is something prior to such a distinction; it is intimately tied to what we ordinarily have in mind whenever we say that we understand a subject.

None of this, however, means to suggest that quantum theory, though bereft of an explanatory ontology, has, as it stands, no physical content whatever. Obviously, the theory must be firmly linked to solid experimental ideas, or else it can predict nothing. Thus, for such quantities as position, momentum, energy, time, mass, charge, and spin as well as notions such as particle and wave, to have physical significance, they must to be tied to measurement. And, of course, measurement is what conveys information to the senses by means of appropriate observational (and usually "amplifying") systems—systems that translate alleged microscopic happenings into the macroscopic ones that experimenters actually observe, grasp, and record.

Indeed, what physicists predict and finally end up seeing as a result of measurement, even in the most arcane quantum experiments, is communicable only in terms of ordinary familiar experiences such as clicks, scintillations, vapor trails, meter readings, patterned shadings on photographic plates, and so on. These are all fully coherent, discrete, and classically describable, familiar experiences. Moreover, as every quantum experimenter knows, the devices (colliders, scatterers, detectors, absorbers, reflectors, computers, etc.) to which these experiences are tied in any quantum measurement are all familiar and accessible things that obey strictly classical principles.

QT, however, does much more than merely correlate raw data via classically rendered measurement setups. The fact that it arose in the very midst of real experimental issues makes it the heir of a rich patrimony of reality (ontological) concepts such as microparticles, spin, waves, and so on. QT, it seems, held to some classical moorings with implicit and often not so implicit references to an underlying subject matter that seemed at least analogous to the original classical one.[4] Indeed, the legacy of classical physics to QT runs even deeper. The quantum-theoretic categories of physical description remain given in terms of both status and change, and these, in turn, are specified in terms of location and momentum—the coordinates that physicists use in classical description. Unsurprisingly, therefore, the expression for the total energy (the "classical Hamiltonian") of a physical system figures centrally in the very formulation we call "quantum mechanics."[5]

But at the same time, QT, with its successful and often surprising predictions, seriously strains the relationship between the classical and quantum worlds. If the theory suggests anything at all about some quantum substrate, what it suggests is not stable and coherent enough to provide a basis for explaining the details of what the theory predicts—certainly not in any philosophically satisfying sense of "explaining." There are too many gaps and too many perplexities.

Indeed, in not providing such a coherent reality model or ontology, QT draws nettlesome ontic and semantical questions from the objective realist sector. This is a sector whose philosophical outlook is the one that still dominates at virtually all levels and branches of natural science. Are there, in fact, such "things" as microparticles and microstates? And how shall we understand pervasive terms such as: "physical wave," "particle," "charge," "field," "empty space," "virtual," "real," "possible," "objective," "determinate," and so on?

The phenomena that QT predicts are paradoxical enough from any classical or even commonsense viewpoint, but the structure and content of the theory itself, as we shall see, presents its own set of enigmas. As a result of both these aspects of QT, the characterization of quantum reality has been and remains a stubbornly resistive issue. This is the issue of interpretation, and it virtually defines the history of the foundations of quantum mechanics.

2

Quantum Puzzles

A Clash with the Traditional Realism of Natural Science

The Core Formalism

The paradoxical features of QT are familiar even to its most casual students, and they have been widely discussed in a vast technical and popular literature that goes back nearly to the first days of the subject. Let us, however, highlight the major ones in order to help fix our ideas and the background of discussion. Consider first the formalism itself of which there are several versions, all demonstrably equivalent. We will refer mainly to two versions, the Schrödinger wave mechanics, inspired by the famous matter-wave hypothesis of Louis de Broglie, and the Heisenberg matrix mechanics that immediately preceded it in the first days of the subject. A combination of these two formalisms constitutes everyday quantum mechanics, which we refer to in our discussion as QT.[1]

The Schrödinger formalism is the so-called formal bread and butter of QT and is by far the most suggestive one from the viewpoint of interpretation. It consists of two parts: a deterministic one serving as the "core formalism" and a probabilistic (nondeterministic) one supplementing the core formalism. The deterministic part of the formalism consists of a set of "quantum rules" that feature the quantum mechanical state function, also called "the quantum mechanical wave function," "the Ψ-function," or simply, "Ψ."

The burning question, right from the beginning, has been: Does Ψ represent anything physical? It is of course a mathematical function—one that evolves with time and whose temporal behavior is governed by the time-dependent "Schrödinger wave equation" in which it is embedded. This equation is the partial differential, wave-type equation familiar to every elementary student of quantum mechanics.

It is homogeneous (i.e., every term in the equation contains the Ψ-function) and linear (neither Ψ nor any of its derivatives occurs to any power greater than one).[2] Ψ as some sort of wave function has amplitude, phase or state of development, and frequency. Its amplitude may even have directional (polarization) properties,[3] all of which are very suggestive of physical waves. But Ψ also has some rather abstract features. To begin with, it is inherently complex, not real.[4] But more than this, it "moves" or, better, evolves *temporally* in an abstract Hilbert space, a "configuration space" whose dimensionality is three times the number of particles in the particular system under consideration.[5] Still, Ψ, in some unclear sense, is understood to "designate" or "stand for," an evolving physical system. This flatly "physical" interpretation, however, raises some pointed questions about the kind of reality Ψ is supposed to represent.

De Broglie and Schrödinger took the formalism literally, and doggedly wanted to coordinate real "material waves" with the Ψ-function. Unfortunately, however, such "waves" cannot possibly be detected in any physically satisfactory sense of detection. The Ψ-amplitudes are not force-linked displacements in any force-stressed environment or field as are, for example, the amplitudes of mechanical and electromagnetic waves. Consequently, they "carry" no kind of measurable "wave-energy." Also adding to the conceptual difficulties is the fact that the formalism requires different wave forms for different observables. Position, for example, requires a pulse; linear momentum calls for sinusoidal waves; and spin employs spherical harmonics.

The Hypostatization of the Core Formalism

Worth noting in this general regard, is a tendency in foundational discussions to proclaim the materiality of whatever Ψ designates on the basis of the fact that the core formalism is genuinely *dynamical;* that is, it moves deterministically with time. Accordingly, this core formalism consisting of the time-dependent partial differential equation, often referred to as the equation that governs the evolution of Ψ, is thought to be strongly analogous to the equations of motion for a classical system.[6] This, it is claimed, is indicative of the substantiality of what is being "propagated." Moreover this substantiality is what interacts (and this can only mean causally) with the measurement apparatus and eventually with the observer, resulting in what is actually observed.

Indeed, given the appropriate initial conditions at any time and place, Ψ is completely determined *mathematically* for any future time and place. What is also true, however, is that the purely algebraic behavior of the time-dependent function, Ψ, that is, its computability or "determinateness," does not, by itself, establish any ontological reference. To say that it does is to commit to a bluntly gratuitous hypostatization that, despite the beguiling suggestiveness of Ψ as a time-dependent mathematical function, confuses mathematical with physical existence. Until we decide, therefore, what sort of thing 'Ψ' designates, the notion of any "causal" interaction between a cluster of "mathematical possibilities" and an observational setup remains utterly obscure. And even more obscure is the notion of an interaction "reducing" whatever Ψ is supposed to represent to the macrophenomenon we actually experience. The gratuitous reification of purely mathematical relationships therefore must be judged unavailing in any attempt to attribute explanatory content to QT.

The Nondeterministic Part of the Quantum Formalism

Hardly six months after Schrödinger published his famous paper, Max Born initiated the probabilistic part of the formalism with his probability interpretation of Ψ. For a single particle, he construed Ψ to be the amplitude of a wave whose intensity, $|\Psi(x,y,z)|^2$, at x,y,z, gives the probability (understood as a relative frequency over a large number of similar measurements) of finding the particle at x,y,z.[7] Generalized for more complex systems, this interpretation, now almost universally accepted, construes the evolving wave function as a propagation of probabilities, thus lending strength to the notion that what is going on is something "wavelike." What, then, is this "physical probability" that evolves temporally as a property of matter-waves?

Wave-Particle Duality in Quantum Experiments

Things would seem strange enough if the quantum formalism hinted that quantum reality consisted of waves and nothing else. The strangeness, however, plunges to bottomless depths when this same formalism, via its predictions, offers hints that these waves are particles as well. As every quantum experimenter knows, and as the formalism correctly predicts, what one always detects in quantum measurements are indeed particles. Wave aspects appear only under certain very special experimental preparations and solely in the distribution patterns that individual particles make at the detection apparatus.

The most frequently cited scenario for depicting the elusive wave-particle nature of the electron is the double slit experiment—an analogue of the historically famous classical experiment. with the same name, in physical optics. The experimental setup essentially consists of a source of electrons that move in the direction of an obstructing screen in which are two closely spaced parallel slits so that some electrons can go through the slits. At some distance beyond this obstructing screen we have a receiving screen that is capable of recording any electrons that get through the slits and strike this second screen.

Classically, one would expect any electron to strike the screen at a place that lies in a straight line connecting the electron source and one of the slits. QT, however, predicts, and experiment confirms, that any electron passing through any slit will, at random, strike the receiving screen at one or another of a variety of (possibly) other places. More particularly, the electron is expected to and in fact will strike the screen at any one of the places within the outlines of the double slit interference pattern. Indeed, if a succession of single electrons were sent through the slits, they would, one at a time, form a pattern of hits on the screen resembling what one would get if a wave had gone through both slits on the way to the receiving screen. And this, of course means that there will be places on the screen that no electron will strike. Apparently, some attribute of each electron—something wavelike—has been propagated from the source, *through both slits*, and onto the screen. Some mysterious "aspect" of the electron, therefore, if not the electron itself, had, somehow, to encounter both slits simultaneously, that is, be in two places at once!

The renowned theoretician and former classmate of mine, Richard Feynman, who never accorded any intuitive coherence to quantum mechanics, embraced this "absurdity" and gave it a formal basis with his "sum-over-paths" formulation. On his account, the electron does actually go through both slits at once. More absurdly still, it simultaneously traverses every possible trajectory on its way to its final destination, the screen. Remarkably, to each of these paths Feynman was able to assign a metric and then calculate a combined average that yielded the results obtainable by the ordinary wave mechanical approach. So, on Feynman's account, it would seem, the electron in such an experiment is not only in two places at once but in an infinite number of places at once!

This of course raises the issue of what one can possibly mean by "single object." Ordinarily for "single objectness," we require single place occupancy at any given time, and as soon as two places are simultaneously occupied we say "two objects" rather than "one." (Try objecting to a he-was-somewhere-else alibi on the basis of denying this common sense rule, in court!) If a quantum double slit experiment indicates that at some instant "the same electron" seems to be in two places at once, one might want to say that at that instant there are two appropriately related electrons rather than one. But then, how did one electron, even for an instant, become two without violating bedrock conservation laws? One could tolerably forfeit particle determinateness and stability by saying that the electron became a wave when it encountered the two open slits, later collapsing back into a particle as it hit the detecting screen. Such a chameleonic duality, however, would be too "ad hoc-ish" both for theoretic comfort and for the general understanding.

At any rate, in this double slit arrangement the electron seems to exhibit *inherent* wave properties. Particle properties, however, also persist. Not only is it always individual electrons that hit the receiving screen, but if we close one slit so that the electron encounters a single open slit, it goes through as a classical particle, hitting the screen without regard for any interference pattern and at a point precisely co-linear with the slit and the electron source. The electron therefore behaves as wave or particle depending on the experimental setup (two slits or one) for observing it.

Some might want to dismiss the wave-particle puzzle by pointing to the unification of wave and matrix mechanics by the von Neumann's Hilbert Space formulation of quantum mechanics, which entailed a shift from the phase space of classical description to an abstract Hilbert space. It has been claimed that this shift involved a transition from standard logic or a "Boolean" algebraic structure to a noncommutative (non-Boolean) one,[8] thus reducing the problem to one of how to make sense of a non-Boolean world.[9] The key notion here is that of "making sense." No doubt, articulating the quantum formalism in non-Boolean, set-theoretic terms can certainly add formal clarity and elegance while of course meeting all consistency criteria, but it still leaves us with wave-particle duality as an ontological puzzle that violates our deepest understanding of what it is to be (physically) a particle. Articulating the quantum formalism in set-theoretic terms (another abstract formalism) takes us no closer to the desired ontological intelligibility. Formal integrity, by itself, cannot yield the *physical* integrity we need for explanatory force, and, on reflection, it seems difficult to deny that explanation, either of a strong (causal) or weaker (unifying) variety, is inseparable from "the making sense of things."

The two-slit scenario, however, presents an even greater perplexity. As the electron starts out from the source and we have blocked one of the slits, how does it "know" that we have done so and that therefore it should start on a perfectly classical pathway through the open slit? Or, consider the following: With one slit blocked we allow the electron to go through the open slit classically so that it may continue on its way to the screen where it is expected to land as a good classical particle directly in line with the open slit and the source. Before it arrives at the screen, however, we open the closed slit. Will the electron now strike the screen in accordance with the one-slit or the two-slit arrangement?

A thought experiment along similar lines, but using photons, is John Wheeler's "delayed choice" experiment, first conceived hypothetically in a purely theoretical context and later performed actually. Consider the pathway of a single photon from an emitting source to the slits and beyond. To intercept the photon after it has encountered the slits, we can place, at some distance past them, two detectors each of which is aimed in direct line with a different one of the slits and the electron source. If we do this, we can expect (and experiment confirms) that one or the other of the detectors will click, indicating that it has registered a photon. Our expectation that this will happen is grounded in the quantum-theoretic assumption that detecting the photon by direct aim at the slits along one or the other pathway causes a collapse of the wave function of the entire experimental system, so that no wave characteristics will occur. Moreover, by noting which detector clicks we can tell through which slit the photon has passed and hence which path it has taken on its way to the detector. In such a case, the photon has behaved as a bona fide particle.

We can, however, prepare an alternative setup for intercepting the photon after it has encountered and passed the slits. We can focus it onto a photosensitive screen. With this second setup, QT predicts and experiment confirms that an interference pattern will appear on the screen signaling that the photon has behaved as a wave and has therefore "gone"—as we might say—through both slits. Modern measurement techniques with beam splitters and mirrors enable the experimenter to switch from one of these experimental setups to the other within the ultrashort time it takes the photon to go from the slits to the end of its journey (at a detector or at a screen). The choice of setup can therefore be made within this ultra-micro time interval.

When such a "delayed choice" is made, we get a bewildering result. The path apparently taken by the photon—one slit or "both slits"—seems to depend entirely on how we choose to detect it, even if the choice is made after the photon has already passed beyond the slits! What we do *after* the photon has encountered and passed the slits seems to determine the very nature of that encounter, that is, whether the photon has gone through one slit (therefore behaving as a particle) or both slits (therefore behaving as a wave). Astounding! An action at a given time seems to have "caused" a state of affairs "existing" at an earlier time. Moreover, once again, we have the dependence of either wave or particle manifestations on the manner of experimental observation.

In a cosmic version of Wheeler's thought experiment, light from a source billions of light years away arrives here on earth after passing by a galaxy or other massive object also billions of light years away.[10] The massive object (e.g., a galaxy) intervenes in the line of sight between the earth and the distant source. The gravitational field of the massive object acts as a lens and bends the light as it passes along the two possible pathways, one on either side of the massive object. The two bent pathways therefore intersect before reaching the earth and form two separate images of the light source as viewed from earth with two telescopes aimed respectively at either side of the massive object. In such a case, each photon has behaved as a particle that has hurtled through space on one or the other side of the massive object. If, on the other hand, we observe the photons by focusing them on a detecting screen we should expect, and in fact do get, an interference pattern suggesting that each photon has split in wavelike fashion and taken both pathways at once past the intervening massive object. Once again, therefore, QT tests the limits of intelligibility by suggesting that a choice we *now* make for detecting a photon from the distant source determines what that photon did billions of years ago! It was quantum weirdness of this sort that led Feynman, Bohr, and other theoreticians to say that no one "really" understands QT. What they had in mind was no doubt the strange, if not obscure, nature of what QT, with its experimentally correct predictions, seems to be saying about the dynamics of an underlying microreality.

Particle-wave duality is the earliest noted and perhaps most celebrated curiosity of quantum phenomena. Not only is the formalism indicative of it, but Niels Bohr enshrined it in his principle of complementarity, which formally acknowledges the dual aspects of quantum reality.[11] The propensity of quantum subject matter to manifest this duality is a basic starting point for all subsequent attempts at assigning a coherent ontology (i.e., interpretation) to QT.

The question of whether a physical entity is a wave or a particle is a familiar one that virtually defines the history of optics. In the centuries-old polemics surrounding that issue, it is generally presupposed that an entity cannot have both wave and particle properties. Indeed, nineteenth-century classical physics, which culminates in Maxwell's electromagnetic theory, keeps the concepts of wave and particle entirely disjoined thus rendering it logically impossible, on pain of contradiction, to attribute both wave and particle properties to a single entity. QT calls for loosening the two concepts so as to make the dual ascription in some sense possible. The result is a conceptual (semantic) liberalization with profound epistemic implications. For one thing, it presupposes that we can somehow know and "understand" such a dual ascription.

This epistemological broadening, however, lacking the halo of familiarity and common sense, puts severe stress on our intuition. The usual way of responding is that of imagining an otherwise nondescript substrate, namely, "wavicles" with *dispositional* wave or particle properties, depending for their actualization on the experimental setup. This kind of ontological retreat, however, still leaves us wondering about any coherent dynamical structures of wavicles that might explain their dual dispositions.

Superposition

By far, the most challenging and complicating feature of the quantum formalism, from the viewpoint of interpretation, is *superposition*—a feature that results from the very form of the Schrödinger equation, namely, that it is a "homogeneous, linear, partial, differential equation."[12] As a result of this formal feature, if the wave functions, Ψ_1 and Ψ_2 represent possible energy states of a system (i.e., particular solutions of the Schrödinger equation) then so does a linear superposition of these, that is, the more general solution, $c_1\Psi_1+c_2\Psi_2$. A most general solution of the Schrödinger equation is therefore $\Psi= \sum_i c_i\Psi_i$ where the Ψ_i are linearly independent orthogonal eigenstates of some dynamical variable, the c_i are complex coefficients, and the summation over i is a summation over all allowable states of the system.[13]

This means that the time-dependent wave equation for a quantum system is understood to represent not only a denumerably infinite superposition of complex functions evolving in a Hilbert space but also a propagation in real space-time of possible outcomes of measurement.[14] For any given time and place, this allows an unlimited array of such outcomes each having a calculable and determinate Born probability, and any of which—it turns out—may randomly "materialize" on the occasion of measurement.[15]

In this scenario a problem of consistency arises. Consider an aggregate, Q, of many particles each of which is capable of having one or another of two incompatible attributes, X and Y. Let Ψ_1 be a state in which all the particles of Q have X and Ψ_2 be a state in which they all have Y. Now let Ψ be a superposition of Ψ_1 and Ψ_2. Paradoxically, QT allows the possibility that for the superposed state, Ψ, (1) 70 percent of the particles will have property, X, and, at the same time, (2) 70 percent will have property, Y. This of course implies an overlap in which the particles would (simultaneously) each have both X and Y, which are assumed to be incompatible.

To resolve the inconsistency, a sizable portion of theorists would appeal to a widely accepted interpretation of QT known as the Copenhagen interpretation according to which neither X nor Y exist except when they are detected by direct measurement. Further, they will maintain, we can test for (1) or for (2), but not for both. For in testing for any one of these we would incorrigibly disturb Ψ making it impossible to test for the other. Moreover, only the measured one of the two attributes comes into existence at the time of measurement.[16]

The strangeness of QT deepens further with the strictly formal fact that any quantum state is itself representable by a vector that can be expanded, as a "superposition of states." This expansion can be made in many ways depending on the "basis" chosen for the vector space. What is generally referred to as "a basis" is a set of state vectors satisfying certain algebraic requirements that relate to their abstract "orientation" in a multidimensional Hilbert space and to the probabilities associated with the components in the superposition.[17] The monumental problem arising at this point, however, is the fact that the Schrödinger formalism provides no guidance whatever for making such a choice.[18]

The Measurement Problem

In the context of a "superposed" formal framework, a very serious problem arises the moment we attempt to say what is actually going on *physically*. To begin with: What (if any) metaphysical status do superpositions have? Is a superposition merely a formal stage of information in a purely computational process, or does it exist as some sort of physical state—impossible though it may be to imagine what it would be like to observe or experience it? Or, again, is it only its separate components that exist, and if so, in what sense of "existence" and how do they become observable on the occasion of measurement?[19]

What we directly observe in any act of measurement is only one or another component of a superposition, and then, of course, not until the measurement is made. Prior to the direct realization of some one component through measurement, therefore, the superposition components seem to "exist" as merely *potential* rather than *actual* alternative outcomes of measurement.[20] Further, with regard to the measurement process itself, the puzzle deepens much more to become the notoriously resistive issue known as the *macro-objectification problem* or, more simply, *the measurement problem*. This problem is one of the most central in the philosophy of science and has incessantly rattled the foundations of modern physical theory since the very dawn of quantum mechanics.

One way of stating the measurement problem is this: the formal evolution of Ψ leaves us with superposed states (eigenstates) that somehow "co-subsist" at most as a set of possible measurement experiences. Setting aside the issue, however, of whether the components of this superposition are no more than a set of merely mathematical alternatives, or else, ontologically genuine actualities evolving with time, an urgent question arises: How is it that on the occasion of observation, what we always see is a "single" and definite outcome? How does the indefinite multiplicity delivered by the formalism become the definite "unicity" we experience whenever we make an observation?[21]

The system, S, for example, may be an electron prepared for a measurement of its spin. Its physical state with respect to spin is described quantum mechanically as a superposition (sum) of the two states, spin of +1/2 and spin of −1/2 (along some given axis). If we allow that the measuring device, M, is also a quantum system capable of interacting with S, then, as a result of such an interaction we have a composite system now quantum mechanically describable as a superposition of two composite states—one in which the detector has coupled to the state of S and points to spin +1/2 and the other in which the detector has coupled to the state of S and points to spin −1/2. Now, if this is the composite quantum state just before observation, how is it that when the observation is finally made, the observer always sees one and only *one* definite pointer reading? (What is more, all other observers corroborate the definiteness). Indeed, we cannot even imagine what it would be like to have an incorrigibly direct "experience" of the superposition that subsists immediately before measurement.[22] Binocular aberrations such as "seeing double" are of course correctible.

The measurement problem, however, does not end here since other closely related issues arise. Is there, in fact, an interaction between what is being measured and the measuring setup (plus or minus observer)? Is any part of the process statistical (stochastic), and if so, why? Further, if there is an interaction between the measuring entities and those we are measuring, what does it say about the realities and the dynamics involved? For example, to which realm (classical or quantum) does the measuring entity belong? And, in particular, how does the observer or, better still, her/his observational experience fit in? Or, putting it differently, where in the composite of "measuring" and "measured" is the cut between quantum and classical reality to be drawn?

Measurement issues such as these have dogged theoreticians throughout the entire history of the foundations of QT and have generated a wide and sometimes confusing diversity of opinion. Obviously the measurement problem is intimately tied to how one interprets quantum mechanics—itself a vastly complex issue. Indeed, one's explanation of the measurement process will certainly imply something about how one views any of the purported realities involved. And this, as we shall note later, is especially so if the measurement process is to any extent seen as a dynamically interactive one.

The Projection Postulate: Wave Collapse

It was in the face of this conundrum that John von Neumann introduced his famous wave collapse or "projection" postulate as the indeterministic addition to the "core" quantum mechanical formalism. This postulate dovetailed with Max Born's probabilistic interpretation of wave amplitudes, thus completing the second of the two parts of the quantum-theoretic formalism, namely, the probabilistic or indeterministic part. In consequence of the predictive success of the Born postulate and of the explanatory appeal of the von Neumann collapse, both contributions are now generally recognized as part of the formalism of standard quantum mechanics.

Von Neumann begins by assuming that in nature there are two kinds of processes: those of the "first kind" (expressed by the von Neumann-Born part of the formalism), which are instantaneous, irreversible, discontinuous, indeterministic, and noncausal, and those of the "second kind" (expressed by the Schrödinger part of the formalism), which are of the usual continuous, deterministic, and presumably causal sort.

These two kinds of processes are both operative where measurement is concerned. The measuring apparatus, A, though a macro-object, is, for von Neumann, a genuinely quantum entity with which the system, S (an evolving superposition of alternative states) can interact. In the process of measurement, S interacts to form a more complex superposition consisting of S coupled with A. Von Neumann then assumes that with the *conscious* act of measurement (which necessarily includes the act of observation) there is an abrupt (nondurational) change of the first kind in which this superposition of eigenstates collapses ("reduces," "projects") into one distinct state (eigenstate) that actualizes in real space-time for observation. Any one of the "coexistent" branches of the composite superposition may actualize, though the more probable ones will do so more frequently under repeated measurements of the same kind. To calculate the probability that any particular branch of the superposition will actualize we implement Max Born's probability interpretation, which equates that probability with the square of the amplitude of the wave function corresponding to that branch.

This extended formalism in which observables are represented by operators does correctly predict the definite (unitary) nature of what we in fact observe when we make a measurement. Von Neumann's postulate tells us that upon measurement there is a collapse whose result is an eigenstate of the representative operator, and an eigenstate is a state in which the measured quantity has a definite (unitary) value. Also intelligible enough—in purely formal terms—is the notion of a set of logically distinct and mutually incompatible "possibilities" each qualified by a probability. So far so good.

As we reflect further, however, the scheme falters and falls woefully short on explanation. The metaphysical status of the evolving superposition is shrouded in mystery, especially as to how on contact with an observer, it becomes a definite experience. Indeed, QT is entirely silent on explaining how any such "collapse" can happen. In what sense are the "potential states" or "propensities" in a superposition a dynamically evolving reality capable of finally "actualizing" on the macro level? It would seem that the superposed states must, as it were, "coexist" in some inaccessible ontological heaven that keeps them evolving and "saved" until a measurement is eventually made.[23] This potential sort of "subsistence" seems to have no philosophical precedent. Even Aristotle, whose reality dynamics pivoted on distinguishing between potentiality and actuality, regarded these two aspects of being as only relative levels of actual existence. In relation to its antecedent state, each Aristotelian potentiality is a genuinely actualized state which in turn expresses a new potentiality for yet a subsequent actualization.

The welter of divergent opinion following von Neumann's overall take on measurement, however, is a testimony to the strangeness of his postulate. A rule allowing an irreversible, discontinuous, and indeterministic process is, by itself, methodologically uncomfortable enough. But introduced alongside a core dynamics that features linearity, determinacy, and temporal continuity, it seems awkward if not grotesque—especially in the context of a scientific culture that favors the very characteristics which collapse violates. Nor does the awkwardness end there. The extraneous and highly stipulative character together with the indeterminateness of what the rule describes tarnish the cogency of the postulate as an explanatory principle.

Adding to the awkwardness of the projection postulate is the patently vague and infelicitous language in which it is couched. Terms such as "measurement," "measurement apparatus," and even "observation" are not easily accommodated in what is supposed to be a fundamental law of physics. Even if we could say precisely what constitutes a measurement, how do we stipulate precisely where and when it occurs? If the unobserved physical world is all quantum mechanical and evolves deterministically, interacting with "layer after layer" of superposed unobserved physical reality, where does the reduction to a definite eigenstate (Von Neumann's "movable cut") finally occur? And, if this critical event takes place in the observer's body, then precisely *where* in the observer's body?

Von Neumann was aware of this problem and finally concluded that it must occur not in the observer's body but in the observer's consciousness. His thought went roughly as follows: Recall that the measuring apparatus is itself a quantum system, which must be "coupled" with the system being measured to form a composite quantum system of superposed possibility waves. For an observation to occur, an act of measurement would have to occur to cause a collapse or "projection" of this composite, multipossibility system. A conceptual complication, however, arises. The composite measuring-measured system is only a system of "possible outcomes" that must itself be measured in order to become "actual" for observation. But this can be done only by coupling the composite system to yet another measuring system. Again, for an observation to occur, the new composite system would similarly have to be coupled to still another measuring system. There obviously looms here an unending regression that cannot be allowed. The measurement process must have a terminus marked by an observation, and this requires a wave reduction or "collapse." The ever-receding boundary between measuring apparatus and observation led von Neumann to maintain that the reduction can occur only in the observer's consciousness![24]

What to make of von Neumann's conclusion here is a matter of some conjecture. Is the act of observation simply an occasion for *accessing* consciousness which, then, in turn, deludes itself by "experiencing" a definite outcome while in actual, physical reality the underlying state remains an "uncollapsed" superposition? Or does consciousness, in fact, bring about (i.e., cause) the collapse?

The first of our two readings of the von Neumann postulate seems, on the face of it, crudely gratuitous and of course requires no *physical* collapse at all. Taking it seriously, however, seems to entail more unsettling consequences. If the experience of a definite state is only a delusion, then the superposition that is up for observation need, in fact, never collapse at all. We could be left, therefore, with a world of unactualized "observables." We have here what amounts to the characterization of any physical system as a superposition of states that "exist" or, better, subsist *in potentia* with at best a delusion mirroring, rather than actualizing, some one component of the superposition. This, however, still leaves the "delusory" experience of a definite (rather than "superposed") outcome of measurement unexplained. What connection—causal or other—is there between a superposition as a physical state governed by the Schrödinger equation and the illusion of a definite outcome of measurement? Without an answer to this question, we are left with experimental observation as an emergent Berkeleian dream whose "confirmational power" is either totally mysterious or totally nonexistent. (For the desired linkage, Bishop Berkeley resorted to God—an inconceivable stratagem in the present context.)

Our second reading of the von Neumann conclusion avoids some of the difficulties of problematic terms such as "measurement," and "measuring apparatus," by putting the measurement cut at some "causal interface" between the physical and the mental. This, however, presupposes mind-body dualism and an answer to one of the most difficult and persistent issues in the philosophy of mind, namely: *How does one fit the mind-body relation (causal or other) into a naturalistic account of the physical world?* Von Neumann appeals to the fact that philosophers have allowed conscious states to produce physical effects such as my hand going up if I *decide* (conscious act) to raise my hand. But that a physicist—albeit a mathematical one—should uncritically introduce consciousness as a constitutive and causally efficient parameter into a purportedly naturalistic rendering of a physical phenomenon is rather surprising. Could von Neumann have been unaware of the labyrinthine and centuries-old polemics on the relation between consciousness and physical states—part of the so-called hard mind-body problem? How, we can still ask, does consciousness, *by itself*, ever "cause" a measurable dynamical effect in a closed physical system?

Other responses to von Neumann's postulate have attempted to do without collapse altogether. The most well known of these attempts is David Bohm's *hidden variable* theory of half a century ago, which we will take up in chapter 4. More recently, other "no-collapse" interpretations have been formulated, resulting in some rather astounding and belief-straining proposals. These will be considered in chapter 5.

3

More Quantum Puzzles

Quantum Uncertainty

Uncertainty of a theoretically fundamental sort is a signature feature of quantum theory. References to this feature, however, both in popular and textbook discussions *about* quantum theory are often somewhat loose and inexplicit, thus adding to any elusiveness already inherent in the subject. Uncertainty in quantum theory has several aspects which some accounts fail to distinguish carefully or fail to relate properly to each other or to the quantum formalism in general.

Consider first that aspect of quantum uncertainty usually referred to as *the probabilistic nature of quantum theory.* In this regard, it is important to note that there is nothing probabilistic in the Schrödinger equation. This is what we have referred to as the *core formalism* of QT, and it completely determines the evolution of the "wave function," Ψ, in a fully deterministic manner. Remarkably, however, this core formalism, by itself, tells us nothing about where to find a quantum particle until we supplement it with Max Born's probability interpretation of Ψ's amplitude. This is an interpretation suggested by the compelling fact that the patterns we observe experimentally in quantum experiments are patterns one can predict if one assumes, as did Born, that the squared amplitude, $|\Psi(x,y,z,t|^2$, is the probability of detecting a particle at x,y,z,t.

Indeed, Born's rule, together with the core formalism, takes us a step further. The core formalism yields the exact evolution of Ψ as a superposition of state vectors. So that Born's probability interpretation of wave amplitudes then translates into the probability that any one of these component states will materialize upon measurement. This is a probability that may generally (though arguably) be understood statistically as a relative frequency over a large number of cases. It is therefore the Born-enhanced formalism and not the core formalism alone that defines the irreducibly probabilistic nature of quantum theory. This could be seen as an element of overall disunity in QT, so that many notable attempts have been made to address the matter. Most notably, theorists have introduced additional underlying parameters in order to "deepen" the theory so as to take it, so to speak, beneath the statistics and down to the underlying individuals. These attempts, however, face serious difficulties of their own, to which we will give some attention later.[1]

Another aspect of quantum uncertainty is what physicists refer to as indeterminacy attributable to several basic ("canonical") quantities ("variables") such as position, momentum, time, etc. What immediately comes to mind here is, of course, Heisenberg's uncertainty principle, which, as every student of the subject knows, relates uncertainties for certain pairs ("conjugate pairs") of variables.[2] Our consideration at this point in our discussion, however, is the attribution of indeterminacy to an individual variable made without making explicit reference to any pair membership. It is natural, therefore, to ask how such "single variable" references to "indeterminacy" follow from the quantum theoretic formalism. Are they merely a consequence of the uncertainty principle or do they follow from the formalism without explicitly invoking the Heisenberg principle?

In the way of answering—consider, for the purposes of exposition, the location of a quantum particle. The indeterminacy of this quantity, which follows directly from the probabilistic nature of quantum theory, is often noted without making any explicit reference to a second, formally related indeterminacy, namely, that of its conjugate partner, the particle's momentum. This single quantity ascription of indeterminacy is typically made by considering the flight of a quantum particle. In such a case, the wave function,Ψ, represents a wave packet within which the particle may be found. But though the formalism tells us that the particle can be found anywhere in the wave packet where Ψ is not equal to zero, it provides no way of predicting exactly where the particle will be found. In this sense, therefore, its position is left indefinite or *indeterminate*. And, as we may add, this leaves room for Born's probability interpretation, which assigns a probability of $|\Psi|^2(x,y,z,t)$ that the particle will be found at x,y,z,t.

Showing with a general argument the indeterminacy of the momentum of a quantum particle is a formally more subtle matter. We can, however, view the simple case of a particle moving in free space so that its potential energy, V, is zero. For such a case, a solution of the Schrödinger equation says that Ψ is an infinite wave train rather than a wave packet.[3] Specifying the momentum of the particle, however, requires specifying the velocity of the associated wave packet (a group velocity)[4] and this can be done only if there *is* a wave packet (not so in this case). On these considerations, therefore, the momentum of the particle remains entirely undetermined. We can, however, let Ψ range over as large a region as desired with Ψ dropping to zero beyond the chosen limit. In this way, we can approximate as closely as we want to the original infinite wave function. We can then normalize the resulting wave packet so that the appropriate probabilities become applicable.[5]

The wave packet will now have some velocity (group velocity) and therefore momentum, approximating as closely as desired that of the particle. Still, we cannot say just what this momentum approximation is. Any wave packet, however, can be expanded as a sum of wave trains,[6] and here again there is room for a Born-type interpretation. The square of the absolute value of the coefficient of any wave train in the expansion is taken to give the probability that an observation will show the particle to have the velocity and therefore the momentum associated with the corresponding wave train. As in the case of position, the momentum thus remains a matter of ineradicable probability and in this sense, indeterminate. Corresponding analyses can be made for the individual members of other conjugate variable pairs.

So much for the fundamental and uneliminable indefiniteness (uncertainty) of the position and momentum of a quantum particle. Let us now go to the Heisenberg reciprocal relationship between these uncertainties. We can minimize the indefiniteness of position by letting $\Psi = 0$ except in a very small region. But the formalism shows that such a small wave packet, will spread out very rapidly resulting in an indefiniteness in the distance the particle will cover a moment later. But this introduces a corresponding indefiniteness in the particle's position at that later moment and therefore in the velocity or distance covered over that short time. The velocity and therefore the momentum are consequently rendered increasingly indeterminate. Or, putting it differently, the velocity of the wave packet at any instant depends on its rapidly expanding size. In determining such a velocity by taking a number of measurements we would get a widening "spread" or error in the results because of the rapid expansion.

We can, on the other hand, let Ψ have the form, $Ce^{-2\pi i/h\,(Wt-px)}$ over a large but, this time, *limited* region so as to determine the particle's velocity and therefore momentum with some accuracy. If we do this, however, we incur a large indefiniteness in the position because of the large extent or spread of Ψ. Indeed, it can be shown analytically (i.e., as a theorem of the formalism) that, in general, $\Delta p \Delta x \geq h/2\pi$

where Δp and Δx are measures of uncertainty in the position and the momentum respectively.[7]

This is the uncertainty principle—one of the most celebrated and distinguishing features of quantum theory. It applies to and only to those pairs of quantum variables known as *canonically conjugate variables*. These are all pairs of variables whose corresponding algebraic operators do not commute, that is, the order in which they occur when they form a product makes a difference. Examples of such pairs are: position and momentum; energy and time; and the angular position of the momentum vector and corresponding component of the angular momentum.[8] The uncertainty principle tells us that the precision for any variable of a conjugate pair is always attained at the cost of precision in the other. But more than this, the principle also implies the ineradicable indeterminacy of any conjugate variable, i.e., the impossibility of zero error for that variable since it entails a singularity, namely, division by zero.

Apart from the formal-type considerations we have given to quantum uncertainty, various intuitive approaches, not all entirely satisfying, are often invoked to enhance our grasp of this basic quantum feature. On one account, an analogy is drawn between the "matter waves" associated with a microparticle and the diffraction formed by a photon of wavelength, λ. Imagine such a photon moving as a dim collimated beam (i.e., a plane wave) toward an intercepting screen with a narrow slit in it, of width, w. The light will impinge on the slit and produce a narrow beam emanating from the other side of the screen. As the photon goes through the slit, its location relative to the sides of the slit can range anywhere within the quantity, w. This range is therefore an uncertainty of location. Call it Δx. We can reduce the uncertainty by further narrowing the slit. As we do this, however, the beam starts to diffract and spread, thus changing the direction of motion of the photon and therefore the direction of its momentum, p. Momentum, however, is a vector quantity, so that a change in the direction of p is a change in the quantity, p, whose tilt can now range over any angle within the divergent beam.[9] This is therefore an uncertainty in p. Call it Δp. Diffraction theory, however, tells us that, approximately, $\Delta p = p\lambda/w$. But $p = h/\lambda$. (the de Broglie equation). Therefore, $\Delta p = h/w$. We see, then, that while reducing the slit width, w, decreases the uncertainty, Δx, doing so increases Δp. Moreover, since $w = \Delta x$, $\Delta x \Delta p = h$, which, in all essential respects, is the uncertainty principle.[10]

In a more direct approach, a moving particle is quantum mechanically associated with a packet of "matter waves" moving as a group with the particle's velocity.[11] The waves in this packet vary in length, but their average length, λ, is h/p. We may find the particle at any place in the packet where Ψ is different from zero. The more precisely determined a particle's position, the narrower its wave packet. This, however, means fewer waves and therefore less accuracy in the determination of λ. The increased uncertainty in λ of course implies an increased uncertainty in p which equals h/λ.

In still another approach, expositors of indeterminacy often resort to the familiar so-called disturbance model. Accordingly, for precisely locating an electron or other very small particle, we might imagine bouncing a very high frequency photon off it and viewing the rebounding photon with a microscope. The Compton-type impact of the photon will cause an abrupt change in the momentum of the electron and also a range of possible deflections of the photon. Moreover, the change in momentum of the electron will vary with the angle at which the photon is scattered. To narrow down the possible directions of the rebounding photon, we would have to reduce the aperture through which the photon enters the microscope. But this cannot be done without prohibitively reducing the resolving power. We are left therefore with an uneliminable indefiniteness in the momentum of the recoil electron. A detailed analysis can show how these considerations lead to the kind of reciprocal indefiniteness expressed by the Heisenberg uncertainty relation. But though this thought experiment may tweak some visualization of what is supposed to be happening, it leaves indeterminacy basically mysterious because it rests on the Compton Effect, which, on quantum mechanical grounds, can provide only a probability distribution for the possible directions of the scattered photon. Unlike any classical account of collisions, therefore, physical analysis along these lines cannot show how to compensate or "correct" for the purported "measurement disturbance."

In yet another intuition-friendly approach, one can argue for the futility of eluding the uncertainty principle with a thought experiment that attempts to measure both the momentum and position of an electron with indefinitely increasing precision. The idea is to intercept the electron with a single photon of reduced frequency (reduced energy) in order to minimize disturbance of the motion. One does so, however, at a cost. The uncertainty in the electron's position measurement is directly proportional to the wavelength of the photon, which of course increases with reduced frequency. The error in the position measurement therefore increases as we try to reduce the error in the momentum measurement. This simple thought experiment based on a few well-established quantum properties of the microworld lends some reasonableness to the uncertainty principle.[12]

In another measurement-related account we allow a photon to pass through a Fresnel diffracting pinhole of width, w. If we allow light to pass through for a very short time, Δt, then on the far side of the hole we will have a wave packet for which we can give a Fourier analysis. This will represent the frequency, ν, not as a definite value but spread out through a frequency band, $\Delta\nu = \nu/n$ where n is the number of waves in the packet, and $n = c\Delta t/\lambda$. Therefore, $\Delta\nu\Delta t = 1$. But $E = h\nu$ so that $\Delta E = h\Delta\nu$. Therefore, $\Delta E\Delta t = h\Delta\nu/\Delta\nu$ and therefore $\Delta E\Delta t = h$, which, again, in all essential respects, is the uncertainty relation. So if we try to determine precisely when the photon goes through the pinhole, the energy of the photon becomes somewhat indeterminate. Once again, the energy-time uncertainty relation is entirely analogous to the momentum-position uncertainty relation ($\Delta p\Delta q = h$), with energy analogous to momentum and time analogous to position.

Indeterminacy is no doubt one of the most distinguishing features of quantum phenomena. Stemming from the wave-particle description of nature, it is uneliminable and deeply inherent in all quantum-theoretic accounts. Most importantly, it must be absolutely and radically distinguished from uncertainty in classical measurement, the latter being due not to any wave aspect of matter but to the statistically tractable random error in all physical measurement. In classical measurement there is of course unavoidable disturbance of the measured system by the act of measurement. With no matter-wave dualism involved, however, this disturbance can, in principle, be either made negligibly small or corrected for on the basis of the classical dynamics of the measurement interaction.

Indeterminacy, however, is as counterintuitive as it is deeply fundamental. Denying the determinateness of both the position and momentum of a material particle and therefore denying the very possibility of its having a sharply defined trajectory compromises what we ordinarily "understand" by the term *particle*. We recall here the semantical pressures exerted by wave-particle duality. The puzzles of quantum indeterminacy seem similarly to call for some radical semantical and/or epistemological reorientations.

To dispel any thought of drawing an analogy between the statistical aspects of QT and the classical science of statistical mechanics, it is instructive to consider the following contrast between these radically different theoretical contexts: As a strictly classical discipline, statistical mechanics allows the simultaneous assignment of exact values to the canonical or basic variables, momentum and position, for each and every particle in a given system of particles. In general practical terms, however, such determination and assignment are vastly unfeasible. Moreover, predicting the subsequent behavior of each and every *interacting* particle in the system on the basis of the laws of classical mechanics so as to determine its behavior would be unimaginably difficult—indeed, something utterly beyond our present analytical and technical capabilities. In statistical mechanics, therefore, we settle for deriving relationships that predict the gross behavior of the system on the basis of certain simplifying but plausible statistical assumptions about a typical individual.

By contrast, QT fundamentally precludes any of this. To begin with, the quantum measurement of fundamental (conjugate) variables such as position and momentum necessarily requires respectively incompatible experimental setups—a circumstance that renders their simultaneous measurement experimentally impossible. More fundamentally, however, the impossibility of assigning simultaneously exact values to such basic variable pairs is formally implied by QT, thus precluding, in principle, all possibility of ever computing an exact trajectory for any quantum individual in the system. What quantum theory does in general predict exactly, however, is a statistical account, not about a set of quantum particles, but about another sort of "collection," namely, a large number of outcomes of similar measurements on the same kind of quantum system.

Virtual Pair Production

The uncertainty principle allows what, from a classical viewpoint, could be seen as the most law-shattering of all quantum phenomena. This is the spontaneous production of transient (virtual) microparticles, even in classically "empty" space—a phenomenon that challenges, at least for a moment, the venerable principle of mass-energy conservation. To see how this is possible, imagine trying to determine, at some time, t, the absence of energy, E, at some location in a vacuum. For a precise time location, we must narrow down the time interval of the determination. The uncertainty principle, however, tells us that the smaller we make the time interval the more uncertainty we incur in the energy measurement.[13] On the most dominant interpretations of quantum theory, measurability and existence are, in principle, indistinguishable.[14] This then implies the existence of increasing energy fluctuations in the neighborhood of the measurement. For small enough time intervals, these energy fluctuations can be large enough to manifest the creation of mass particles such as electrons.

Conservation of energy, however, which in QT kicks in at the level of short-time averages, requires that the energy which has been "borrowed from the cosmos," as it were, be returned quickly. This can be done only if the spontaneous "creation" is not of single electrons but of electron-positron pairs. These can quickly combine and annihilate each other to give up a photon that then disappears in "repayment" of the borrowed energy. Indeed, the greater the amount of energy borrowed, the more urgent and therefore the more immediate is the repayment. The pairs produced are said to be *virtual*—meaning that they are extremely transient, the ultra-microtime intervals involved being of the order of 10^{-43} seconds.[15]

A similar account can be given for momentum measurements, except that this time it is the location of the measurement that, if made smaller and smaller (more and more precise), results, under the uncertainty principle, in larger and larger momentum fluctuations. According to QT therefore, nature, at sub-microlevels, is a turbulent discontinuous sea of spontaneous creation and abrupt annihilation even in those places that classically were thought empty.[16]

Phase Entanglement

We have still to mention *phase entanglement* or more simply, *entanglement*—thought by some to be a characteristic feature of QT most at odds with the overall orientation of physical science.[17] Indeed, in some respects, some might see the allowed possibility of entanglement as a methodological retreat from the analytic approach to nature set in place by Descartes, Galileo, Newton, and Einstein. The concern would be that, arguably, such a retreat is a serious challenge to causation as standardly understood and to the method of differential analysis (featuring the differential calculus) by which we study a formally (mathematically) limited and isolated part of the world in order to understand all of it.

Of still more concern is the possibility that phase entanglement may be seen as implying "universal entanglement" or the organic unity of the entire universe. And this could be taken to mean that to "really and fully" know anything one would have to know everything. Or, conversely, if one "really knows" anything then one knows everything. At any rate, entanglement is as unsettling as it is intriguing. Indeed, as a profoundly characteristic feature of QT, it haunted Einstein until his death.

The entanglement of two interacting physical systems means that once the entangling interaction has occurred, the systems, even after eventual separation, can no longer be described as before, that is, as strictly and separately independent states. A widely discussed example is one involving two quantum particles that have been put under certain experimental preparations. QT allows these to interact so as to form a "coupled" system in the so-called singlet state, representable by a single complex wave in an abstract (six-dimensional) space. The two particles will then remain coupled, that is, have oppositely correlated properties, as parts of a single state wave no matter how far apart they may later separate. This means, and experiment confirms, that, if measurement of one particle at any time reveals a positive spin component (along some axis) the other, upon measurement at that time, will be found to have the negative (opposite) of that component.

Quantum-Theoretic Description and Completeness

Deeply troubled both by such entanglement and what it broadly implies about reality, Einstein and two younger colleagues (B. Podolsky and N. Rosen) proposed their famous EPR thought experiment in an attempt to establish that QT was flawed—the flaw being what they called "incompleteness."[18] It is remarkable that, even after nearly three-quarters of a century and a vast literature on the subject, the subtleties of this proposal continue to rattle and divide opinion.

The underlying ontological assumptions of the EPR thought experiment were these: (1) If we can without uncertainty predict (determine) the value of a physical quantity, then some sort of reality corresponding to that quantity must exist. That is to say, the physical quantities in point are "real"; (2) For any physical theory to be "complete," the elements of physical reality must all find their counterparts in that theory; that is, they must be represented and accounted for by the formal elements and structures of the theory; (3) What we do to any system at some location cannot *instantaneously* affect a system at another location. EPR then argued that there are physical quantities that can be determined without uncertainty (disturbance) and that are therefore real. QT, however, cannot account for these, and it is therefore incomplete.

The EPR thought experiment starts by imagining a pair of particles, say, particle A and particle B, that have been made to interact and have been sent off in opposite directions. For such a composite system, the total momentum of the two particles, $p_A + p_B$, and their separation, $q_A - q_B$ is imagined to be known precisely.[19] Consequently, measuring the momentum of A makes it possible to determine the momentum of B by subtracting the momentum of A from the total momentum—a quality that is conserved as the two particles move apart.

Further, measuring the position of A allows a similar determination of the position q_B of B by subtracting $q_A - q_B$ from q_A. Moreover, though measuring the position of A will inevitably disturb the momentum measurement of A, by assumption (3), the position measurement of A can be made without disturbing B. In this way, the position and momentum of B can be determined without disturbance and therefore without uncertainty. This, however, is a result that QT prohibits on the basis of the uncertainty principle together with the fact that the operators representing these observables do not commute.[20] QT therefore would seem to imply that the EPR measurements on A must in some "spooky" way instantly disturb B and preclude the simultaneous determination of p_B and q_B without uncertainty. Such a disturbance would of course compromise the entire EPR argument for the existence of uncertainty-free simultaneous values, p_B and q_B.[21] Therefore, either QT is incomplete or one would have to abandon assumption (3) and allow that measuring q_A could instantaneously affect particle B. EPR, however, maintained that no reasonable conception of reality could permit this. So, it opted in favor of (3) and concluded that QT does not provide a complete description of physical reality.[22]

The defenders against the charge of incompleteness were followers of the Copenhagen interpretation of QT, which was led by Bohr and which largely became the orthodox interpretation of QT. Essentially, their rejoinder was that it is meaningless to speak of the existence of physical attributes except as a result of and during *direct* experimental measurement. It could not be validly concluded, therefore, that any reality had been established corresponding to the purportedly determined (but unrealizable) simultaneous values, p_B and q_B.

Some subscribers to the EPR argument, however, countered by extending it as follows: Consider the spins of two electrons that have been prepared to be in the singlet state, that is, to form an entangled system of zero spin or one in which the spins of the two particles are oppositely correlated. Imagine that the electrons are each moving in opposite directions toward detectors A and B, respectively. The detectors can, at random, be set specifically to measure any spin component (let us say, either the x or z component) so as to determine whether the particle has the positive or negative of that component.[23] The probability of detecting a + or – spin component for any setting is demonstrably 50/50. Moreover according to quantum mechanics, a definite value cannot be simultaneously assigned to both the x and z components.

As the theory predicts and experiment confirms, if the particles have been properly prepared, no matter what component the A-detector is arbitrarily set to register, the B-detector, when set for the same component, must (with 100 percent probability) detect the opposite of that component. Therefore, if, for example, the A-detector is set for the x component and registers +x, the B detector, when set for the same component, will surely register –x. What is more, according to the EPR proponents, this is the very spin the electron would have had from the time it was prepared until the time it was measured. Let us assume, however, that instead of setting the B-detector for detecting the x component, the experimenter at that location sets the detector for the z component. Quantum theory says that the detector will register either +z or –z with a 50 percent probability for each—a result that is of course very different from what would, with certainty, have been detected at B if the A-detector had been set for the z component. Since whatever spin the electron registers at B is the spin it got when the singlet pair was first prepared, what is it that "informs" the B electron on how to orient its spin depending on what the experimenter at A decides?

Further, suppose that the experimenter at A decides to set the detector for the x component and gets a +x result. Then the x component of the B electron is definitely known to be –x. If, instead at A, the z component is measured, then the z component at B is similarly known. Moreover, the electron must have had these spin components from the very time of preparation. This is a quantum-theoretic impossibility. Definite x and z components of electron spin are disallowed by quantum theory. Therefore, once again, the choice of measurement at A (i.e., measuring for x or measuring for z) must have an effect at B.

The EPR assumption that the spin of the B electron was an objectively definite spin existing continuously from the time of preparation to the very moment of measurement stemmed from the objective realism in the philosophic orientation of Einstein and his two colleagues. Indeed, he shared this essentially classical outlook with nearly all physicists of the day. The dependence of the outcome at B on the choice of measurement at A, however, left a choice between only two possibilities. Either it meant there had to be some sort of mysterious influence on B by what happened at the A-detector—and, for sufficiently separated particles, the influence in point would have to be transmitted at superluminal speed possibly approaching simultaneity (a somewhat unsettling conclusion in the light of special relativity)[24] or, quantum physics was incomplete because it could not ensure (i.e., provide a coherent account of) the objective reality status of the spin of the B electron until it reached B. Einstein and his colleagues opted for objective realism and therefore concluded that QT was incomplete.

Einstein's problem with instantaneous influences at a distance, also referred to as non-locality or nonseparability of influence, was quite understandable. By allowing instantaneous influences at a distance, QT seemed to threaten one of the most deeply entrenched ontological principles of both common sense and standard science, namely, the requirement that all causes be local (also referred to as separable). This means that any effect caused by an action at some distance must be mediated continuously from point to point along that distance and take a finite (i.e., non-zero) time to do so. As a consequence, no signal from some point, A, can possibly affect any process at point B, if the time required by the process is less than the finite time it takes for the signal to get from A to B.

The ensuing and unrelenting debate between Einstein and Bohr over the issues raised by EPR is a very long and well-known chapter in the history of modern physics. While Einstein held to an objective realist view requiring all physical particles to have actual properties even in the absence of measurement, Bohr allowed quantum particles dispositional or potential properties ("propensities") that could be actualized only by measurement.

At any rate, most notable in this general regard is the fact that the irreconcilable stands taken by these two great physicists were essentially ontological and therefore interpretational in nature. The long debate requires no further review here except to say that J. S. Bell became equally disturbed by the EPR issues and as a result formulated his famous inequality relation, which added yet another complication to the problem of how to interpret QT

Bell's Inequality: The Unyielding Challenge to Traditional Scientific Realism

By the end of the forties, the Einstein-Bohr debate had drifted decidedly in favor of QT. Nevertheless, eight years after Einstein's death, J. S. Bell, still uneasy with the issues raised by the EPR-type thought experiments, proposed one of his own.[25] In his EPR-type setup he imagined pairs of particles, say, protons, prepared ("entangled") all in the same way at some common source, and then allowed to separate in opposite directions. The preparation put the protons in the *singlet state.* This means that the protons of each pair were correlated so that they would be found to have mutually opposed spin components, x or y or z, when each was measured by detectors at opposite distances from the source. At some distance apart, however great, each would be intercepted by a measuring device that could be set at random to measure one or another of the proton's three possible spin components.

It is an experimental fact of particle physics that any spin component can have only one of two opposite values, say, + or –, and that one or another of these two values will turn up randomly for any given component. It is, however, a restriction of QT that, for any one particle, having or not having a spin component can be simultaneously decided for no more than one spin component since the measurement for any one component will disturb the others.[26] This restriction, however, is circumvented in the usual EPR way of making one assignment by direct measurement and the other indirectly. If, for example, one particle of a pair measures spin up (+) for the x component and the other particle measures spin down (–) for the y component, we know that the first particle also has spin up (+) for the y component.

Thus far, Bell's thinking rested on the following assumptions: (1) Any proton can have objective stable attributes independently of whether or not a direct measurement is being made on it (objective realism); (2) No act of measurement at any detector can have any instantaneous influence on any other part of the system. That is, the dynamics involved are strictly local. A simplified version of what Bell then imagined might go as follows: Adjust the detectors separately and at random to measure some one or another possible spin component. Take measurements with the detectors at different settings and consider the number of cases in which the protons of each pair are found to have positive (though different} spin components. Then, on the basis of the above assumptions and elementary statistical considerations, go on to a bit of arithmetic in elementary set theory and derive Bell's now famous inequality which specifies a numerical relationship between sets of protons featuring some combination of different but positive spin components.

Bell's inequality is $n(x^1y^1) \leq n(y^1z^1) + n(x^1z^1)$. In this expression, $n(x^1y^1)$ is the number of proton pairs whose members come up with respectively deduced positive outcomes for differing components, x and y. Further, $n(y^1z^1)$ and $n(x^1z^1)$ are analogous quantities and subject to similar inequalities.[27]

Most remarkably, however, Bell's inequality conflicted significantly with what one would get by a purely quantum-theoretic computation for the same numerical relationship. For, according to QT: $n(x^1y^1) > n(y^1z^1) + n(x^1z^1)$. Obviously, something was wrong either with QT or with the assumptions underlying Bell's thought experiment. If QT was right then at least one of the assumptions on which the Bell inequality is based must be false.

Well, was QT—so uniformly successful in the past—once again right? But then, how could one bet against either the objectivity (i.e., measurement-independence and stability of the proton's properties) or locality so basic to the intelligibility of causal action? It was time—and urgently so—to go to the laboratory to settle the matter this time with some real experiments.

As is well known, this has been done by a good number of investigators, some using protons, others using photons in admissibly analogous ways. Overall, the experimental results clearly favor the quantum-theoretic computation. That is, nature itself violates the Bell inequality exactly as QT predicts.[28] Setting the detectors for different spin components by changing the orientation of one of them apparently influenced measurements at the other enough to change what was expected if no such influence had occurred. Moreover, as experiment showed, this influence had to be instantaneous. The linkage between twin protons was measurably tighter than Bell's inequality would have it. The problem, therefore, was Bell's background assumptions which, as we may recall, were essentially two—a *realistic ontology* on which he based his arithmetical calculations, and *locality,* which ruled out unexpected influences. At least one of these had to fail.

It is difficult for any natural scientist to give up realism altogether. If he/she is to construe science as an activity for informing our understanding of nature, that activity, it seems, must ultimately be a search for explanations in some philosophically satisfying sense of "explanation." And these require an objective, stable ontology for fleshing out an intelligible description of what it is we are theorizing about. Without some such underlying reality framework, scientific knowledge would seem to reduce to a mere set of successful but blind rules for predicting future data on the basis of past or present ones. The scientist's incurable belief in objective substance as the ultimate basis of sense experience (some might prefer to call this "metaphysical wisdom") tells her/him that phenomena alone cannot explain phenomena.[29] The burden of failure of Bell's inequality therefore seems to fall on the assumption of locality. Apparently some phenomena are non-local and, one must say, no less real for it.[30]

It seems appropriate to note here that there is nothing about the non-locality, established by the inequality experiments, that prohibits local causes even within the same system. Indeed the non-local contexts so far considered and observed have occurred within quantum systems whose relevant parts have been prepared (entangled) by means of experimental apparatus and procedures that are quite stable and locally governed. It is unwarranted therefore to conclude that what is required, post-Bell, is a reality model or ontology for physical phenomena in which all local causation is swept aside.

Nor does this mean that if the quantum world—and, by extension, the rest of the world to which it must be naturally linked—is a tapestry of local and non-local causes, those causes are—as one might say—all of comparable "efficiency" (to use an Aristotelian term). As is by now generally recognized, non-local "influences" are severely limited. The experimental violation of Bell's inequality is no more than a correlation of altered random sequences. What the inequality experiments demonstrate for very simple, properly paired systems is that when such systems have been made to interact suitably, they tend to be linked more tightly than one would expect on a strictly local model. What are being observed are random sequences. And what the observed non-local influence changes is not the general fact of randomness but the particular random ordering of one sequence as compared to that of another random sequence. (Clearly, a random sequence can be scrambled and re-scrambled while still remaining random.) Observers at one end of a coupled system, therefore, can receive, instantaneously, no information regarding what is going on at the other. The ordering at their end is not what it would have been had things happened differently at the other end, but they cannot know that anything has changed unless they directly observe the ordering at that other end. Consequently, though QT correctly predicts that there will be non-local influences in certain well-specified quantum contexts, such prediction does not deny the locality of communicable change. *Efficient causation,* as ordinarily construed has not been scrapped.

Still, Bell's discovery is somewhat disquieting from the viewpoint of assigning an intelligible objective ontology to QT. Influences-at-a-distance, if one may call them so, fall entirely outside the familiar web of causal chains that give structure to the world of standard science and common sense. And if this does not shatter our intuition of causation entirely, it certainly does perturb it. A non-local influence is, after all, an action at a distance in the sense of a change being occasioned instantaneously by something happening somewhere else. It therefore rudely violates the mediating time-space contiguities that make local influences so fathomable.

True, quantum non-locality communicates no information and therefore, very strictly speaking, remains consistent with the proscriptions of special relativity on superluminal communication. Such non-locality, however, does pose a discomfiting and head-scratching—though intriguing—background for those relativistic proscriptions. What is more, the thought that the "big bang" may have "entangled" the entire universe can raise haunting issues about the kind of significance to attach to the possibility of a universal entanglement requiring non-local correlations at overarching cosmic levels.

So once again, QT predicts—and does so correctly—another perplexing feature of physical reality thereby further burdening the quest for an ontology that can serve the requirements of an acceptable interpretation. Moreover, indications are that in any such ontology, commonsense realism would have to give way to a more critical realism that allows, at least, for non-locality.

4

Interpretation

The Problem of Interpretation: Old and Persistent

A natural response to the baffling aspects of quantum physics is the demand for an interpretation of QT. The varying and conflicting attempts to meet this demand constitute the problem of interpretation. This is a problem that has occupied theoreticians from the earliest days of the "quantum age" to the present. More recently, it has led to several highly imaginative interpretations by D. Bohm, H. Everett, J. Wheeler, and others some of which we will touch on later.

The problem of interpretation can be expressed as the following question: What can we plausibly say about both QT and the real world to explain how a somewhat freakish and highly abstract formalism can yield its outrageously strange yet stunningly correct predictions?[1] The many attempts to answer this gnawing question have led to a variety of proposals, none of them meeting with sufficient acceptance to settle the issue. Moreover, as we have already noted, the diversity of opinion reflects profound and perhaps irreconcilable differences of philosophical orientation underlying any attempt at interpretation or even some fundamental resistivity of the quantum core formalism to any explanation-giving interpretation.

Objective Realism in Physical Science

It seems safe to say that physicists, like all other students of nature, tend to be incurable objective realists who cling to the rock-solid ontology they inherit from common sense. On every indication and especially in their less critical or philosophical moments, they believe with all their hearts that—whether we exist or not, know it or not, are observing or not, like it or not—there is one objective world of material entities, "out there," whose existence is independent of our own and even antedates it. Moreover, as the belief goes, these are entities some of whose "essential" attributes, both inherent and relational, can at least in principle be discovered and, in some acceptable epistemological sense, *grasped*. Further, any system comprised of such entities is assumed to be, at any time, in a definite state that, in final analysis, depends only on the full physical history of the system and on the universal laws that have governed its evolution.

Of course, no segment of the scientific community will deny that whatever we can know about the world must, in some ways, depend on the nature and limitations of our perceptual faculties and on the human manner of observing and studying. But then, the pervasive, underlying realism sees this only as an epistemic and not an ontological limitation. Moreover, whatever the limitations, science invariably takes its most confident entries to be objective "knowledge of what exists and of how what exists behaves." And though it justifies these entries with observational data, their existential content is neither about the data nor, fundamentally, about when or whether an observation was made.

Thus, investigators report—quite categorically—that the cosmos is fifteen billion years old, or that a known star at the edge of our galaxy exploded eleven billion years ago, or that there was a Mesozoic period ending about seventy million years ago during which dinosaurs ruled the planet, or that there are X-rays ricocheting around the universe from independently existing black holes, exploding stars, and very hot intergalactic gases, or that there are such things as leptons which may be either electrons, muons, tauons, or neutrinos, or that the human genome consists of three billion DNA basic pairs shared by one hundred thousand genes on twenty-three chromosomes, or that there are several hundred billion neurons in the human brain with countless possible networks among these—and so on and on and on. A fortiori, none of these claims is ever said to be true when and only when measurements were, are, or will be made, or if and only if certain kinds (rather than other kinds) of measurements are made, or if and only if our sensory endowment is what it is (rather than different).

We have here a prevailing article of scientific and commonsense dogma that might be seen as an *observer-related symmetry principle.* True, when they are thinking critically and philosophically, quantum physicists such as Niels Bohr, John von Neumann, and many others resort to interpretations, for instance, the Copenhagen interpretation (to which we will turn shortly), that introduce severe existential instabilities as well as mind and observation dependencies into quantum subject matter. All philosophical excursions aside, and when all is said and done, however, empirical science characteristically returns to its core belief in a ready-made world whose existence and most fundamental attributes are ultimately invariant to scientific probing. And, at least in principle, whatever disturbance is caused by measurement may either be made negligible or be accounted (corrected) for, so as to render a result that is, with more or less approximation, true of the undisturbed quantity. Proceeding in this way, therefore, scientific knowledge is thought to converge to an ever-accumulating set of "objective truths" about an "as is," observer-independent world.

What is more, a world seen in this way is an essentially commonsense world. In such a world, an entity, at any given time, can have one and only one set of well-defined dynamical attributes, and all changes of these attributes are space-time continuous. These changes are completely determined by immediately antecedent dynamical states of the entity and its causal or other unifying environment. Some changes are advisably taken to be spontaneous and treatable stochastically—but only for convenience or while we are agnostic about their causes. Further, where changes are to be "causally" related to actions at some distance, the changes must be mediated by intervening chains of actions—chains that propagate at finite speeds and whose links are space-time contiguous. No event, therefore, can ever be influenced by any distant event before the finite time it takes for the influence to arrive. Within that time interval, the event is said to be local, that is, confined to its own "causal neighborhood" or locality. In these precise senses, then, the world of the objective realist (physicist or not) is both entirely objective and entirely local.[2] And, no doubt, it is easily conceded that indeed this is the belief orientation that obviously rules when it comes to preparing and carrying out actual physical experiments.

Closely tied to the objective realist orientation is the widely held though often unstated belief that the space-time structures and movements of actual physical systems make a good fit with our geometric intuition. So that ultimately whatever happens in the real world can be correctly represented in terms of intuition-friendly, three-dimensional, Euclidean constructions. (Time is easily accommodated as a "fourth parameter.") From this sort of belief orientation, it is only a small psychological leap to the widely held view that the mathematical form of a theory as structured, either algebraically or geometrically, is "literally" a key to the structure and movement of the reality it is supposed to explain, or even only predict. This is invariably true of how we "read" the laws of mechanics, optics, electromagnetic theory, and even relativity.

The Schrödinger equation at the core of the quantum mechanical formalism is a "wave equation" in the sense that some of its solutions can be graphically represented either as standing or advancing waves. Indeed, this famous equation is readily extracted from the dynamical context of a vibrating string. Some objective realistic theorists such as de Broglie and Schrödinger in the early years of QT therefore interpreted its formalism literally as representing real waves that actually collapse to yield stable results at the moment of measurement. As we have noted, however, this literal kind of interpretation faced serious conceptual difficulties. A deeper approach was needed, and the early attempts to meet this need resulted in what are known as the hidden variable interpretations (HVI) of QT.

Hidden Variable Interpretations: Highly Problematic

The citadel of objective realism, as we have just outlined it, is classical physics. But some of the most highly distinguished architects of QT have, even in their more critical and philosophical moments, been believers and therefore very hesitant about any physical theory—QT included—unless it presents a world that accords with an objective realist ontology. Only then, according to objective realism, does a theory yield fully intelligible "gapless" explanations of how the natural world actually works.

Accordingly, theorists of the objective realist persuasion have seen QT as flawed in foundational respects. Recall, for example, EPR's objective realist charge that QT is incomplete, because, unless it allows mysterious non-local influences, it fails to encompass determinate and stable aspects of reality. To this line of criticism there is usually added the observation that QT is only a theory of large numbers or aggregates—a theory that provides no more than statistical knowledge of probable states under given initial and boundary conditions. QT, therefore, tells us what *can* happen, what will *probably* happen, but not what will *actually* happen. This, then, gives no determinate account of the actual individual players in physical systems and therefore of the specific nature of physical reality.

Worse still—as the criticism goes—while QT provides statistical predictions, events that may, to some extent, betray some "weirdness" in the underlying reality, it fails to explain them. Indeed, its rules renounce causality and continuity of motion and, in doing so, they forfeit all possibility of providing any intelligible causal account of quantum phenomena. Therefore, it is advised, theorists should attempt to extend and deepen QT. And, to do this, additional underlying entities and processes or "hidden variables" may have to be posited that exist at deeper-than-quantum levels and that may therefore lie beyond the reach of all presently known measurement possibilities.[3] A leading hidden variable theorist is David Bohm, who maintains that just as we explain the motion of Brownian particles by going to the underlying motion of smaller particles at the atomic level, we should be able to explain what happens at the atomic (quantum) level by going to a "deeper"(sub-quantum) level. That is, we should be able "to go to some as yet unknown deeper level which has the same relationship to the atomic level as the atomic level has to the Brownian motion."[4]

He rejects as circular what he considers to be the usual interpretation of QT: "The usual interpretation of quantum theory requires us to renounce the concepts of causality and continuity of motion at all possible levels."[5]

And again:

> It would appear that the conclusions concerning the need to give up causality, continuity of motion, and the objective reality of individual micro-objects have been too hasty. For it is quite possible that while the quantum theory, and with it the indeterminacy principle, are valid to a very high degree of approximation in a certain domain, they both cease to have relevance in new domains below that in which the current theory is applicable. Thus the conclusion that there is no deeper level of causally determined motion is just a piece of circular reasoning, since it will follow only if we assume before hand that no such level exists.[6]

It is worth noting here that on views such as Bohm's, the uncertainty principle is incompatible with causation. At least some physical properties on the sub-quantum level must be free of the uncertainty principle; for if they were not, they could not be simultaneously determined without uncertainty. Lacking such determinacy, the entities involved could not interact deterministically and thus provide the ontological basis for an intelligible explanation of quantum phenomena. More precisely, indeterminate values of "hidden variables" could not serve as the initial conditions that, together with appropriate causal laws, are required for causal explanations.[7]

Theoreticians who like Bohm share the need to commit quantum theory to some sort of "deeper" ontology constitute a group that includes M. Planck, M. Born,[8] D. Bohm, de Broglie, Einstein,

Schrödinger, and a good many others. And their varying approaches to the problem of what to do with QT have come to be known collectively as *the hidden variable interpretation(s)* (HVI) of QT.

From time to time, quantum theorists have either postulated or simply upheld the existence of some hypothetical process or quantity in order to account for one or another quantum phenomenon. They have done this, however, in a very limited way and without changing or else reinterpreting the core formalism of QT. A well-cited example of this is the famous von Neumann projection postulate already discussed in the section on the measurement problem. Another is the objective realist assumption that the indirectly measured observables in an EPR experiment are ongoing real quantities, objectively there, even when no measurement is going on.

HVI, however, is, generally speaking, a systematic program of much more ambitious scope. It attempts to formulate a "deeper" composite "sub-quantum" theory (H) with an ontology rich enough to serve as a causal basis for explaining the entire quantum realm, that is, the predictions of QT—however puzzling they may be—and how such predictions are possible. For causal intelligibility, such an ontology would have to be one that features determinacy of subject matter, causality, and continuity of change. And this could well require introducing entities that, though having full existential status, are "hidden" in the sense of not being within the purview of quantum mechanics and also, possibly, hidden in the sense of being unobservable in present-day experimental contexts. Rather than being only about aggregates, H would therefore be about individuals to which determinate attributes would be assignable and which would evolve causally as definite, objective, individual states under new and possibly very different kinds of sub-quantum laws. A modified form (most-likely nonlinear) of the Schrödinger equation could then be formulated and assumed to be an approximation holding only at the quantum level. The state-function,Ψ, would then represent an average of sub-quantum level quantities, and quantum statistics (which are ultimately about a collection of events) would then be recoverable by some sort of averaging operation over specifiable states of individual systems. Hidden variable theories, therefore, attempt to recover both some aspect of the quantum formalism at the quantum level and all of quantum statistics at the observational level.

In all of this, some modifications of the present-day quantum formalism, such as the addition of nonlinear features, could be introduced to yield the required quantum atomicity or "lumpiness." The modifications could also accommodate certain aspects of the experimental environment and possibly even the history of previous measurements—all of which may be relevant and dynamically inseparable from the total system. In such a framework and with fully determinate initial values of all variables involved, quantum phenomena would, as a good approximation, be both deducible and explainable within a dynamics of individual, "sub-quantum" variables in H. In this sense, QT could then be said to be logically imbedded in H. What is more, hidden variable theorists might hope that such a theory, if appropriately and fully developed, could, as a crowning achievement, consolidate quantum and classical theory into one unified physics of nature.[9]

HVI along such lines as these would of course face familiar philosophical concerns regarding the metaphysical status and physical significance of its "deeper" variables. It would also expect the resulting framework to fulfill its promise to remedy the gaps and perplexities of QT without contradicting any quantum-theoretic result. Moreover, it would have to do so along objective realist lines, which generally require local causality together with the definiteness, stability, and objectivity of the most fundamental postulated entities. This would of course mean no dependence of existential status on what any human or other organism happens to be perceptually aware of. That is to say, the subject matter of any hidden variable framework must be specifiable without reference to anything subjective, such as consciousness, in any of its epistemic states. The hidden variable program however, has spawned an array of issues ranging from the methodological viability of HVI (in terms of physical significance) to others about the very possibility of HVI. And the resulting polemics comprise a hefty portion of the twentieth-century literature in the foundations of QT.

The immediate objection to HVI has been the methodological one that points to the observational inaccessibility of hidden variables. This is an inaccessibility that hidden variable theorists do not see as a serious kind of "methodological" deficit. If hidden variables can both help explain phenomena and lead to correct predictions then their physical legitimacy is to be considered assured. But many theorists—especially those of the generally prevailing Copenhagen persuasion (to be discussed shortly)—are troubled by the "hidden" nature of the strange new variables that HVI postulates. Such theorists—strongly positivistic in their orientation—prefer living with causally unexplained quantum effects over accepting explanations based on hidden mechanisms, which they regard as unnecessary "metaphysical baggage."[10]

Reflection, however, adds other concerns of a less philosophical and more logical sort. These are concerns having to do with the overall integrity and, ultimately, plausibility of HVI. In final analysis, a physical theory is a product of diagnostic reasoning. It is a hypothesis or set of hypotheses (among an indefinite number of possible alternatives) designed to account for (explain) some set of material facts (laws, data, etc.). Even before any sort of experimental confirmation has taken place, some preliminary judgments can be made about how "good" a theory is. These are plausibility judgments based on what one might call theoretic virtue into which go such features as the simplicity of the theory, its type-similarity or other compatibility relationship to the background of established knowledge, and the possibilities it offers for eventual observational confirmation or falsification. On all of these HVI seems wanting.

The most noted attempts at formulating a hidden variable theory are, again, due to David Bohn, and one of his later ones provides some good illustrative material. He assumes that "connected" with each of the fundamental particles of physics there is a very small particle that is always accompanied by a "companion wave," the two coexisting as distinct entities at the sub-quantum level. Bohm further assumes that the companion wave consists of "a new kind of oscillating force field" that satisfies Schrödinger's equation. This means that, rather than being just a formal computational ("algorithmic") device, the wave function, Ψ, actually designates an objectively real oscillating field which exerts a new kind of "force" that guides the particle over a randomly tortuous but definite trajectory. Under the action of this force, the particle generally tends to locate where Ψ has the highest intensity. The same particle, however, also suffers other motions of a problematic origin that are random and oppose this general tendency and result in a wide range of random fluctuations.

Overall, this model is believed sufficient for yielding the results of QT, however puzzling they may seem. For example, it means to make sense of the wave-particle *dispositional aspects* of quantum reality. These "aspects" are reduced, on the sub-quantum level, to the combined physical action of actual waves and actual particles, which explain respectively appropriate parts of what happens experimentally.

So, in the double slit experiment with electrons, the interfering wavelets emerging from the slits guide the particles that "ride" them to where they strike the screen, as particles, to form an interference-type distribution. On the other hand, when one hole is covered, there is only one guiding wavelet and therefore no interference. The interference pattern therefore disappears as expected. Meanwhile, the random fluctuations in the location of the particle provide the stochastic element needed to yield the Born probability postulate and recover all of quantum statistics, thus explaining the random (though patterned) landings on the screen.[11] What is more, the same random fluctuations can be strong enough to explain the strange quantum phenomenon of "tunneling" through a potential barrier, a phenomenon that is impossible on classical terms.[12]

Our representative example of HVI, however, strongly highlights the plausibility concerns we have already mentioned. It suggests that the hidden variable program requires the positing of entities (bodies, fields, forces, processes) and laws that are at once very many, very strange, and very elusive. If simplicity is a signature of truth, then the lapse of frugality in the sheer number of assumptions casts a rather dark shadow on plausibility. Further deepening this kind of concern is the utter strangeness ("newness") of the posited entities, and their relationships raise questions of "fit" with the kind of physical knowledge we already have. Just what kind of sub-quantum coupling is it, for example, that makes a companion wave absolutely inseparable from the particle it guides? Similarly mysterious is the kind of force that the wave exerts on the particle. What kind of law does the resulting interaction obey and how does the sub-quantum force fit in with the four well-known forces of the standard model?[13] A connection between the quantum and sub-quantum levels is established with the assumption that the behaviors of variables on the quantum (i.e., atomic) level are merely averages of behaviors of variables on the sub-quantum level. Apart from this connection, however, there is no observational effect of the sub-quantum level on the atomic (quantum) one. The sub-quantum variables, because of their nature and the sub-quantum laws that govern them, are so new and so radically different as to seem observationally inaccessible under all the known laws that govern any presently conceivable experimental context.

The question, then, is: How great can this strangeness be without seriously compromising the plausibility of HVI? Surely, the greater the strangeness the less the plausibility. Indeed, given the many implicit and exotic aspects of some hidden variables, one cannot even rule out the possibility that any conceivable experiment for confirming their existence would require experimental circumstances and procedures that, on the basis of what is presently known, are physically impossible. And this of course could raise serious questions of consistency with some of our most soundly entrenched laws and theories.

In principle, however, these criticisms of HVI do not explicitly foreclose the logical possibility of observational confirmation for the existence of its "new kinds of variables." It may be argued that the history of reductive scientific theory is a history of variables that, for a time, have in some sense been hidden. The sub-quantum variables of HVI, however, are not just hidden; they seem to be very deeply hidden, and this has been dramatically indicated by assaults on the very possibility of the hidden variable objective.

The history of these objections is a long and convoluted one. To summarize them briefly, we can start by recalling that HVI wants no truck with non-locality. A serious concern of HVI theorists therefore is the finding on the part of some investigators that HVI involves wave collapse. Besides being unobservable, wave collapse is also abrupt (discontinuous) and non-local. These are features that clash head-on with the objective realist orientation that motivates the hidden variable program in the first place. The abruptness of wave collapse allows no time interval within which to fit any standard causal account of what is happening behind the reductive process, thus leaving quantum measurement essentially unexplained in any "standardly" causal sense. The non-locality arises from the notion of an "event" in which the act of observation instantaneously influences (collapses) all sectors of the expanding wave no matter how far apart these might be. In the case of a spin measurement on any one of two electrons in the singlet (entangled) state, for example, the collapse would project the electrons into opposite spin states, and thus, simultaneously involve both electrons, no matter how cosmically separated. On all standard considerations, however, this could not possibly happen if the influence of the collapse at one electron were to propagate toward the other at any finite speed. Any hidden variable account in terms of a wave collapse, therefore, seems inassimilable to a realistic HVI.

In this respect, Bohm's companion waves fare rather badly. Apart from the fact that their energies are so low they cannot be experimentally detected, they also fall subject to non-local influences, i.e., to instantaneous effects in all parts of a wave no matter how distant these parts are from each other. Such waves therefore incur, not only the positivistic scruple against nonobservability, but also the standard realistic objections to non-locality.

Non-locality is an embarrassment to HVI, but most embarrassing of all is the finding that no HVI is even possible without introducing non-locality. The earliest impossibility argument against hidden variables was launched by von Neumann, who first asked: Is any HVI for quantum theory even possible? He began by showing that in order for any hidden variable theory to yield all the results of QT, it would have to require the precise and simultaneous specification of noncommuting quantities such as position and momentum, thus contradicting the uncertainty principle. He then went on to prove the general impossibility of positing any set of hidden parameters for which one could determine the simultaneous results of measuring two noncommuting variables. He did this by showing that doing so would be inconsistent with the rules for calculating quantum mechanical probabilities—rules that are decisively confirmed by experience. In short, the thrust of von Neumann's proof was that the existence of hidden variables was incompatible with the truth of quantum mechanics. Accordingly, if such entities existed quantum mechanics would have to be factually false.

This famous impossibility proof, however, did not settle the matter. One of the postulates in the proof fell into serious question, drawing the charge that the postulate rendered von Neumann's reasoning circular. Bohm, for example, charges von Neumann with improperly assuming that certain features of nature associated with the current formulation of QT are absolute and final at all levels of existence. So that, in order to specify the state of any physical system, one must do so, at least in part, under the rules of QT. This assumption, Bohm maintains, dogmatically sidesteps the possibility that as we go to the subquantum level, the rules that govern quantum observables break down to be replaced by very different ones.[14] The polemics that followed resulted in a long series of works technically critical of one another, pro and con HVI—some of these offering refinements of previous impossibility proofs, others formulating impossibility proofs of their own, and still others objecting to either of these.

Several decades after the appearance of von Neumann's proof, J. M. Jauch and C. Piron used a propositional calculus or "logic of quantum mechanics" that had been developed by von Neumann, to strengthen the proof.[15] They did this by sidestepping the troublesome postulate and substituting a weaker one so as to avoid the charge of circularity.[16] More specifically, they showed that introducing hidden variables into QT would entirely preclude the non-commutativity of variables. That some dynamical variables do not commute, however, is both a fact of experience and implied by QT. Hidden variables therefore are not only incompatible with QT, their existence is unsupported by experience. The authors therefore conclude that "the possibility of the existence of hidden variables is to be decided in the negative."[17] Given the cogency of this argument, the possibility of a hidden variable theory explaining QT, at least in its present form, seems seriously undermined if not entirely foreclosed.

The impossibility charge against HVI was later strengthened by J. S. Bell and several successors who developed proofs—so-called no-go theorems—to show that no *local* hidden variable theory could possibly reproduce the quantum statistics and thus account for (i.e., explain) all the predictions of QT.[18] Moreover this impossibility entails the impossibility of explaining, on the basis of causally independent (i.e., local) hidden variables, certain surprising correlations between spin measurements on two electrons in the singlet state and moving apart in opposite directions. The hidden variables would have to behave as non-local parameters.[19] We are of course referring here to the violations of Bell's inequality both by QT and by experiment.

Still, fresh questions popped up. Investigators soon pointed out that Bell had failed to consider that the measuring process could conceivably affect the distribution of the hidden variables. This, it was alleged, was a possibility that could not be excluded a priori. Encouraged by this criticism, David Bohm attempted to show the possibility of hidden variables by formulating an HV account of quantum measurement on a simple quantum system.[20]

Bohm did this, however, at the price of breaking with one of the strongest formal traditions of physical theory—a tradition relating to simplicity or, perhaps better, to mathematical manageability. For the purpose of relating the state vector, $|\Psi>$ (representing the state of the quantum system), to his postulated hidden variable, $|\lambda>$, he introduced a *nonlinear term* into the Schrödinger formalism.[21] In such a framework, the results of measurement were determined by the initial values of $|\Psi>$ and $|\lambda>$ and the reduction of the wave packet was thereby described as a deterministic physical process. Within such a context, together with some questionable assumptions relating to the random distribution of $|\lambda>$, Bohm was able to recover quantum statistics and reproduce the known predictions of QT. This approach is generally referred to as "Bohmian mechanics."

It may also be said of Bohm's formalism that—as a further complication—it also included mathematical representations that depend on the effects of the specific experimental environment in which the measurement is made. He further proposed a generalization of his equations to accommodate the effects of the values of $|\Psi>$ at every other point in space and to include even effects of all previous measurements (measurement history) on the system. But alas, this holistic representation of measurement makes Bohm's theory non-local and therefore self-defeating from the viewpoint of causal explanation in any traditional sense. It thus puts it at odds with the standard no-action-at-a-distance causality of objective realism and, of course, with relativity as well. Also notable is the fact that Bohm's non-local concept fuses the measured-measuring (composite) system into a "whole" that allows only measurements that reduce the wave function of the "total" system while disallowing this for any component system. This limitation quashes an important possible mode of proto-theoretical thinking that could eventually be fruitful in this unsettled area. It precludes any thought experiment (as for example of the EPR type) for envisioning measurement results on "limited" component systems.[22]

Much more recently, some authors have pointed out that the von Neumann Hilbert space description of a physical system entails the transition from a Boolean to a non-Boolean algebraic structure for the attributes of physical systems. Moreover, the non-Boolean structure is *strongly* non-Boolean, meaning by this that it cannot be extended to a Boolean (classical) structure by adding to that structure other determinate dynamical variables (hidden or not). This precludes the possibility of averaging over the determinate values of such variables in order to recover the quantum statistics.[23] Obviously, this weighs heavily against any hidden variable program.

The desire for a successful hidden variable extension of QT theory has yet to be entirely quelled, but half a century of critical thought would seem to dim the chances of a generally acceptable formulation. What is more, a good many theorists view hidden variable models as hostages to an objective realism that resorts to inaccessible metaphysical baggage. Such models are seen as ontologically inflationary and formally inelegant. Accordingly, many have been drawn to alternatives that entail radical and even surprising departures from the objective realism that has traditionally biased natural science. The dominant historical move along these lines is the Copenhagen Interpretation.

Copenhagenism: No Less Problematic

Historically, the major alternative to HVI has been a set of related approaches which, for convenience, we refer to as the *Copenhagen interpretation* (CI). In its most sweeping, but widely accepted version—strongly tinged with positivism—CI assumes that quantum observables such as position, momentum, and spin, "actually" exist only during measurement. As inter-measurement parameters, such "quantities" are to be banished like the sub-quantum particles, companion waves, and other unobservables that HVI invokes for making sense of QT. As a consequence of these proscriptions, however, CI finds itself with an ontology so severely diminished as to have to settle for merely predicting rather than explaining the vexing yet compelling results of quantum measurement. And, if this ontological thinning leaves us with something vastly counterintuitive about any remaining underlying reality, the apparent "weirdness" must then be granted as irreducibly brute fact.

Still, CI does ease some quantum perplexities relating to particle-wave duality. Recall that QT rules out the possibility of measuring noncommuting variables simultaneously with one experiment. They must be measured successively with respectively appropriate and incompatible experimental setups. This, Copenhagenists maintain, suggests that wave and particle aspects of quantum systems are successive and "complementary," rather than simultaneous and incompatible. Until measured, they are not "actual aspects" but only "potentialities" (dispositions, propensities) of quantum entities to be actualized exclusively under respectively incompatible and successive experimental preparations. What were contrary properties thus become "differing" dispositions dependent for their actualizations on differing types of experimental contexts.

The obvious infelicity in this is that it shifts the wave-particle mystery to the level of dispositions. What sort of "entity" is it whose particle properties can "dissolve" into those of a "wave" with an altering of experimental setup? By contrast, the relativistic dissolution of mass under nuclear compaction or other exothermal processes is rendered coherent with mass-energy equivalence and the consolidation of our conservation laws to accommodate the conservation of a composite "mass-energy" substance. The dissolution of particles into "matter waves," however, offers no comparable possibility, there being no specifiable dynamical substrate for linking the two "reality forms." Prior to measurement, therefore, "whatever" is evolving deterministically under the nomological constraints of the Schrödinger equation seems to have no more "being" than that of a sort of "virtual reality." And this would seem to consist of a blend of incommensurate *potentiae* that co-subsist and co-evolve in one and the same "virtual system," if we may call it such.

Common sense, however, asks: How does such a shadowy being manifest the "causal bounce," as it were, to evolve from moment to moment and then, only during a measurement, to *actualize* as some detectable wave or particle property depending on the nature of the measurement? Further, the "phenomenon," or whatever it is that is going on, leaves us clueless as to how the system under measurement "knows" what the setup is. By what interactive mechanism would an electron, for example, become "geared" to exhibit the appropriate property?

Further distancing itself from the objective realism of both HVI and classical theory, CI with its measurement-dependent evanescent ontology is able to dismiss the disturbance model of measurement. A later measurement cannot of course disturb the outcome of an earlier one if that outcome no longer exists. The disturbance model—kept alive in HVI—is thus rendered dispensable. This result is especially satisfactory for those quantum theorists who have qualms about the essentially classical nature of the disturbance model and therefore about its legitimacy as an explanatory account of any quantum mechanical feature.

At this point it might be noted that Copenhagenists are sometimes characterized as hard-nosed positivists who renounce all physical ontologies for QT except sense data. Indeed, in denying existential status to unobserved variables, Copenhagenists do reveal a strong positivistic leaning. But flatly characterizing them as positivists would be far less than accurate. Every indication (including the language Copenhagenists use) has it that CI subscribes to an ontology of genuine particles, spins, locations, quantum states, etc.—diffusively indeterminate, existentially dependent on observation, two-faceted, and otherwise incoherent though they may be. CI also accepts the prevailing probabilistic rendering, due to Max Born, according to which the square of the Schrödinger wave amplitude represents the probabilities of certain measurement outcomes and according to which "probability waves" evolve in space-time as dynamical, though mysteriously undetectable, realities. CI theorists, therefore, though less vulnerable than their hidden variable colleagues, are not entirely out of the range of any extremist positivistic charge of metaphysical turpitude.

Most unfortunately of all, however, CI leaves quantum measurement itself no better explained than it is by the HV models. Indeed, von Neumann wave collapses are themselves hidden mechanisms. Moreover, in the final analysis, the CI measurement model leaves the *macroscopic* measuring devices, as "causes" of wave collapse, entirely outside of quantum systems and therefore of quantum physics. The measuring world is allowed to remain solid, stable, and entirely classical, while the measured quantum world is not even there until the very act of detection! And even then, CI requires that any measurement outcome—indeed, any experience relating to the microworld must itself be described in fundamentally classical terms rather than as a quantum state. The "two worlds" are therefore left entirely disjoined yet permitted to "touch" during obscure and irreducible measurement interactions that abruptly and mysteriously create or annihilate observables! The awkwardness of this "macro-micro" divide shades into a serious theoretical question: Where, within the "micro-meso-macro" spectrum, do we locate the divide for distinguishing, say, a multiparticle quantum state from a classical one? It may be further noted that CI—its indefinite and observation dependent *micro*-observables blinking in and out of existence—demurs on the issue of why *macro*-observables invariably retain their sharpness and objective stability. All told, there is here an ontological gap that CI leaves unattended and that seems to foreclose the possibility of any full recoverance of macrophysics from quantum physics.

In the way of "softening" this metaphysical dualism, some writers of standard texts often make reference to *Bohr's correspondence principle* as a basis for seeing a measure of unity between quantum and classical physics. Doing so, however, overstates the case for the theoretical content of this important principle, which is no more than a rule or guide for making quantum-theoretic choices. The correspondence principle stipulates that as we approach certain limiting conditions in a given bounded space, quantum results must approach those of classical mechanics. This is a rule that has proved useful for guiding the formulation and development of QT. Shall we, for example, let Ψ be complex? Yes, if doing so assures correspondence.[24]

And, as might well be expected, the application of Bohr's rule has so tightly guided quantum theorizing that the resulting quantum formalism does indeed reveal the required correspondence. Thus, for any particle with momentum, p, moving toward an obstacle of linear dimension, x, as the product, xp, becomes very large compared to Planck's constant, h, the particle's wave aspects start to vanish, and it starts behaving classically. It of course follows that, in a world where h were to be zero, or the wavelength of the particle, λ (= h/p) became negligibly small, the classical laws would hold at all levels of physical existence.[25]

But while Bohr's principle has been immensely and very pointedly useful especially in developing some of the rules (e.g., the selection rules) of quantum mechanics, it is entirely unavailing on the issue of theoretical comprehensiveness. The formal, and experimentally well-documented transition from one conceptual framework to another as we go to the limiting conditions, does not quite provide the desired unity. In this actual world, $h \neq 0$. Quantum laws, therefore, are distinctively and radically different from classical ones and yield dramatically different predictions not only for the micro-dimensions that define the "deeper" side or "substrate" of physical reality but also for the macro-phenomena that defied classical explanation. The historical failures of classical theory attest well to this, and what persists, therefore, is a reality sundered in two at the measuring-measured interface, with each side of the divide under radically separate laws.

Indeed, though the predictions of QT are classically surprising and astoundingly accurate, when all is said and done "quantum mechanically," what particle physicists have ever observed are events (scintillations, clicks, traces) that, together with the instruments of experimentation, always play out macroscopically, in the laboratory procedures, very much according to classical laws. And, neither the literalists like those of the early years, nor the later hidden variable theorists, nor the Copenhagenists that challenged both have been able to bridge this deep chasm in a generally satisfying way.

Finally, CI leaves totally unaddressed the charge that QT is unacceptably "incomplete" in at least two senses: First, with the existence of its entities dependent on whether or not an observation is going on, CI leaves no place for the stable inter-measurement reality (Einstein called it "interphenomena") that the objective realist tradition in physical science is loathe to relinquish. Second, what CI finally provides is not an account of individuals with determinate dynamical attributes but a statistics of aggregates, that is, of multiple events and multiple alternative outcomes.

5

Fresh Starts

The Many Worlds Interpretation: New and Old Problems

The past four or five decades have seen a number of theorists making fresh attempts at interpreting quantum mechanics. The aim in these attempts is to address the issues left behind by CI, the most pressing ones being the "micro-macro" dualism in modern physics, the closely related measurement problem, the recovering (explaining) of quantum statistics, and the salvaging of some sort of realism without incurring the pitfalls of hidden variables. Preferably, this is to be done without wave collapse, hidden variables, or Born's probability interpretation. The key point of departure for these theorists, however, has generally been the issue of superposition and measurement, which they take to be most central to the overall problem of interpretation.

Superposition is dealt with by positing a universe of many worlds in which each of the possibilities in a superposition is a genuinely existing reality in a different world.[1] This means that all the physically possible outcomes of measurement delivered by the core formalism are to be considered actual and respectively observable in one or another of many worlds—each home to one or another state of the observer.[2]

We have here a full rejection of both wave collapse, hidden variables, and the probability interpretation of Ψ. Wave collapse is seen as patently unobservable, and non-local, as failing on the measurement problem and, to boot, as a discontinuous assumed "process" nowhere implied by the core or basic Schrödinger formalism. It is therefore unwanted as sorely extraneous and "ad-hoc-ish." Further, it is observationally inaccessible and therefore metaphysically burdensome. And, as shown by Bell's work, it is, like all other hidden variables, unavoidably non-local. Similarly, the Born probability interpretation, also nowhere implied by the core formalism, is taken to be another add-on that could be dispensed with by means of, possibly, a more encompassing account.[3]

But despite the "thumbs down" on both the von Neumann and Born postulates, as well as on HVI, these current theorists are not, by any measure, ontological minimalists. They part radically from any positivistic influences and shrug off all "verificational" admonitions by positing some of the most imaginative and expansive—if not extravagant—ontologies in the history of modern physical theory. H. Everett in 1957 (later with J. Wheeler), for example, came up with the *many worlds interpretation* (MWI) of quantum mechanics,[4] and a substantial number of differing variations have been elaborated since. Most generally, however, MWI construes the total universe as one entangled system of subsystems of subsystems, etc. This "quantum totality" or "maxiworld" encompasses the entire universe of entities and processes to include conscious observers, measuring devices, and everything else. The time-dependent Schrödinger equation governs this maxiworld, which accordingly evolves smoothly and deterministically as a vast, all-encompassing superposition of superpositions.

One version of MWI, arguably the most dominant and not clearly Everett's, has the evolving maxiworld actually splitting and differentiating into distinct branches ("worlds," "parallel universes"). All the states of this "total world" are superposed states with components that are, in fact, actual evolving physical states rather than vaguely construed "evolving possibilities" waiting to be "collapsed" and thus actualized, by eventual measurement, into discrete and definitely observable realities. With such a "solid" thriving ontology, MWI ostensibly has no need of extraneous collapse postulates or of the foggy notion of ghostly "potentialities" "languishing," as it were, to be made observable by measurement.

When a measurement finally occurs, it results in a composite (entangled) physical system consisting of what is being measured plus the measuring device plus the observer's state of consciousness, which is itself physical and therefore able to become entangled as part of the evolving composite system. This composite system, then, is what finally evolves as a superposition of all the physically possible (alternative) outcomes of the measurement (i.e., all those with probability greater than zero). Moreover, each of the components or "branches" in the "decomposition" of the superposition is an actual rather than merely a potential outcome of the measurement. Indeed, what seems to be at play here is a physical pan-realism that encompasses and reifies all but the physically impossible.

The awareness or conscious state of the observer will of course be different in each branch of the decomposition, since it is assumed to "reflect" the state of the measured system and of the measuring device (pointer reading) in that particular branch. Each of the resulting multiplicity of conscious states may thus be thought of as defining a so-called "relative state" or remainder of the branch consisting of the branch minus the conscious state of the observer in that branch. Further, with entanglement considered pervasive, MWI sees each conscious state of the observer as defining a relative state consisting not simply of the branch in which it finds itself but of the rest of the entire universe. The composite of the observer's consciousness and the cosmic remainder would then be a "current" world, that is, the world that the observer, in his particular conscious state is actually aware of.[5] It is further assumed, however, that the observer in any of his alternative conscious states is strictly and necessarily unaware of the other universes, each of which is home to an alternative outcome of measurement inclusive of a corresponding alternative state of the observer's consciousness.

MWI in its various versions has enjoyed considerable attention since it was first proposed. But it has also generated much contemporary discourse on deeply challenging issues. To wit, the MWI ontology is a vast one, elaborate enough to invite the charge of compromised simplicity when compared to, say, wave collapse. What is more, nearly all of this posited reality is necessarily and therefore eternally inaccessible to any current observer, thus marking MWI as an Alice in Wonderland dream construed merely for the sake of overall theoretic smoothness. More pointedly, however, our culturally engrained "positivistic" restraints, with which we generally test ontological limits, can arouse here as much concern about "physical significance" as would any hidden variable or wave collapse process. Myriad, undetectable, "psychocentric" other worlds would seem especially irksome in this regard. Finally, we might add that there is some mystery about the splitting of the maxiworld into distinct parallel worlds—as posited in dominant versions of MWI. What are the underlying mechanics, and when or how does the splitting occur? The dynamical elaborations this calls for go far beyond what any current version of MWI offers.

Of the interpretational tasks that face MWI on the ground, however, perhaps the most immediate one is that of accounting for the statistical aspects of quantum phenomena—aspects that, as we have already noted, the core Schrödinger formalism does not deliver. In addressing this matter, MWI starts with its fundamental assumption that on the occasion of measurement, the composite system, consisting of that which is being measured, the measuring device, and a brain branches out into a superposition of alternative outcomes each with an alternative brain and an associated mind or observer that "experiences" outcomes. The amplitude of the evolving superposed wave function does, of course, suggestively vary in the region of each alternative outcome. MWI, however, rejects the Born probability postulate as formally extraneous. It therefore cannot use the wave amplitude to assign probabilities to the alternatives.

To provide the needed probabilities, some theorists call for vastly expanding the already bloated ontology of MWI to the extent of assuming that the set of many worlds is, at all times, a *continuous infinite* set.[6] This allows the introduction of a measure defined on subsets of these worlds that satisfies the axioms of the probability calculus. Further, the evolution of the universal state vector is assumed to be such that the measures of the resulting subsets will correspond to the frequencies we actually observe in quantum measurements and which, of course, are the very frequencies predicted by the Born postulate. In effect, it is simply *supposed* here that what one might call the "sizes" of the resulting subsets are directly proportional to the probabilities (correctly) predicted for alternative measurement outcomes in standard QT. This means that, on the occasion of measurement, the proportion of worlds that evolve into states corresponding, respectively, to the various superposed outcomes of a measurement, mirror the observed quantum probabilities (frequencies) of the outcomes. To paraphrase Lockwood, every observation in a given branch or "world" is thus regarded as a cookie jar sampling of the total "world of worlds" represented by the evolving universal state vector.[7] This "measured" ontology is therefore posited diagnostically, precisely for the purpose of explaining the empirical findings resulting from the "samplings."[8]

This interpretation thus purports to offer a customized ontology for recovering quantum probabilities in an allegedly "natural manner."[9] But what can the term "natural" here convey more than the notion of mathematical admissibility? From the viewpoint of theoretical elegance, this expanded MWI ontology seems, in final analysis, no less arbitrary and extraneous to the basic framework than the von Neumann measurement postulate. Indeed, we can assign measures to the spaces in many other ways, all satisfying the probability calculus. And there is nothing in the core formalism or in any other available framework to indicate which measure to adopt for making correct predictions.

Less urgent physically but more intriguing metaphysically is another issue posed by the branching model of MWI. As has been noted by D. J. Chalmers, MWI says nothing about how to account for our very strong sense of personal identity.[10] I am surely the very same individual who was working on this manuscript last week and who began writing it three years ago—and who once married a lovely poet named Gaetana. But am I the only consciousness associated exclusively with one of the many brain states that have branched out successively in their evolution to the present? And what about the consciousnesses in the other parallel worlds. Given the branching model of MWI, wouldn't their identities as persons also thread back to some shared past?

In this regard, I may be willing to grant that though the other selves in the multiverse are inaccessible, their sharing a portion of my past seems to compromise neither their personhood nor my own—at least as sensed by me in my own particular world. But a universe of parallel worlds and selves (persons) that lie beyond all physical possibility of experiential realization is, again, a belief-straining ontological curiosity. And the apparent extravagance of it all is especially highlighted by the supposition that these vast stretches of purported reality are, in principle, not only experimentally inaccessible (in any direct sense), but also eternally inconsequential.

For the issue that most deeply afflicts MWI, however, we must return to the problem of measurement. This is the lingering and dogging mystery of an evolving quantum reality that is irreducibly superposed but that, on every occasion of measurement, appears not as some incoherent jumble of alternatives, but as the definite (discrete, unitary) world with which I ("this" observer) am currently associated, The alternatives featured by any such evolving superposition represent experiences that are mutually exclusive, at least from any ordinary phenomenological viewpoint. We know, however, that our observations are definite (discrete, coherent) experiences rather than unimaginably multiple jumbles. And, as we also know, this "unicity" meets with general intersubjective agreement.

MWI, as we have noted, responds to this with its maxiworld ontology. In doing so, however, it suffers yet another complication due to the purely formal properties of the abstract space quantum that state vectors inhabit. Any state evolving in accordance with the Schrödinger formalism can be decomposed as a superposition in an endless number of ways depending on the *vector basis* chosen for the decomposition, since in any world there is a mathematical infinity of such bases to choose from.[11] Well, then, how is any sort of "choice" made?

Faced with this challenge, the many worlds theorist assumes that for each of the parallel worlds there is, as a matter of "natural fact," a so-called preferred basis in a Hilbert space that makes possible a decomposition each of whose components features only definite ("unitary") experiences for the particular consciousness that figures in the component. This, then, is the basis the observer in each world in fact "chooses" from an indefinite number of possible ones for the decomposition of his world. The resulting decomposition defines a class of definite alternative outcomes of measurement and therefore of actual alternative worlds, each "housing" a corresponding state of the observer's consciousness. In this way, one and only one of the alternative states of the outcome indicator (dial, pointer, etc.) figures in any actual experience. The result is a multiplicity of worlds each with a different conscious state of the observer experiencing a different but discretely definite world. Moreover, each observer state is utterly incognizant of other worlds and of the observer states they house.

Cool! But a rankling question persists known as the "preferred basis problem." The existence of a "natural basis" which in some sense is "preferred" and which accordingly the observer "chooses" is in no way required (implied) by the Schrödinger formalism. Indeed, the formalism provides no guidance whatever along these lines. Therefore, any metaphysical privilege or special appeal that might be accorded some particular basis would be just an arbitrary assumption reflecting, rather than explaining, how we actually experience the world. The response to this difficulty is a set of interpretations that are closely related to MWI and that are collectively known as the many minds interpretation (MMI) of QT.

The Many Minds Interpretation: More Problems

MMI rejects altogether the notion of an objectively preferred basis, noting that the laws of physics provide no such privileged metaphysics. It therefore accepts the account that the core formalism of QT gives of the physical world. And, like MWI, it accepts the "multiverse" of superposed physical states that accordingly evolves smoothly and deterministically. But it goes farther: without assuming the existence of any preferred basis, it associates with any observer, at any given time, a simultaneous multiplicity of points of view or "minds," each having its own experiential content. This idea is first introduced by H. D. Zeh, which he calls a "many consciousnesses interpretation."[12] The idea, however, is explicitly developed by D. Albert and B. Loewer who first call it the "Many Minds Interpretation."[13] In dropping the troublesome assumption of a preferred basis, however, this interpretation must expand the vast ontology of MWI even more by greatly multiplying the number of evolving worlds. For a state superposed with respect to one basis is also a state superposed with respect to all others, and the number of these is infinite!

At any rate, this gives us a multiuniverse or multiverse that evolves seamlessly and totally as a vastly superposed state with none of the splittings or differentiations often associated with MWI. The splittings and branchings in MMI are of minds or "conscious points of view" rather than of worlds (MWI). Moreover, the division or differentiation occurs only when an observation is made, at which moment there is a differentiation within the observer's consciousness into alternative viewpoints (perceptions, experiences). Thus, at the end of a measurement, according to MMI, there is a whole set of alternative states of observer awareness—not the one conscious state that, on MWI, would have been made definite (singularized) by some hypothetical preferred basis.

But alas, this still leaves us asking the ultimate measurement question: Why, in the context of multiple mind states, is the observer's actual experience of a measurement outcome always discrete (singular, coherent) and, as such, intersubjectively corroborated? The response of the many minds theorist is most remarkable: the experience of definiteness is only an appearance, not an ontological reality. As such it is an illusion of consciousness—a singular limitation, "quirk," or even "prejudice of the subjective." It is an illusion even though the intersubjective corroboration of this "illusion" is unexceptionable! Well, then, what are we to make of this illusion, and how does it arise? The MMI reply: explaining why such an illusion should ever occur is not a task for QT. It is the burden of some other theory (if one should ever become available) about how "conscious mentality" relates to the physical world that quantum mechanics presents.[14]

Referring to illusion and then outsourcing the task of relating it to the rest of reality is nothing short of astounding. It is a blunt dismissal of the measurement problem by "kicking" the discreteness (unicity, definiteness, coherence) of observational experience "upstairs" to the metaphysically vague realm of illusion. Such a move, however, is too facile a sidestep to take seriously. It severely snags the phenomenology of experience and the role of observational confirmation (or falsification) in physical inquiry. Indeed, it is the stubborn stability of experimental or even commonsense observation that serves as the court of ultimate appeal for establishing any kind of reality (i.e., matter of fact) claim. In this regard, illusion guarantees nothing. And, if observational experience as we know it is some kind of illusion, then at least in this particular or "current" world, it must be a coherent and totally shared one so that it may ground all our reasonable expectations and practical beliefs.

The retreat to illusion on the part of MMI is hardly a move toward explaining measurement as a window of access to physical nature. But whether or not one can make sense, in terms of illusion, of what we actually experience during measurement, leaving the underlying psychophysical relations to some other (possibly future) theory is a serious omission. It leads to an unsettling dogmatism regarding the ontological status and causal efficacy of conscious states or, more generally, of mind. Once consciousness (illusory or not) is brought into any physical theory, its role in the dynamics of the theory must be addressed. And this unavoidably requires some sort of commitment on the most fundamental issue in the philosophy of mind, namely, the issue of how mentality relates to at least some other things in the world. Disappointingly, many mind theorists tend to make the same kind of uncritical leap in this regard as do their many world colleagues. The trip plate is the mind-body relation, and inevitably, they must choose between describing it dualistically or materialistically. This is a choice that cannot be narrowly guided by the particular needs of a physical theory, to the utter neglect of other considerations—considerations usually developed in more critical philosophical contexts and in the cognitive sciences.

In this regard, the offhand assumption of MMI, inherited from MWI, is that conscious states (which one can take to be the same as mental states) are physical. This, however, stirs up a concern properly attributed to D. Albert and B. Loewer by M. Lockwood.[15] If conscious states are physical then according to QT they can, like all other physical states, be superposed. But the idea of superposed mental states (e.g., beliefs or memories) makes no sense. A mind cannot be regarded as superposed. If it were, it could not, without further assumptions, be considered accessible by introspection. And we know that such mental states as recollections of measurement results are in fact accessible by introspection, which reveals them to be of a definite rather than jumbled, superposed nature. In a modified version of MMI, therefore, Albert and Loewer opt for dualism by taking minds to be "something other than physical." Indeed, if minds cannot be superposed they must a fortiori fall decisively outside the reach of the quantum mechanical formalism according to which the entire physical world is subject to superposition.[16]

As in the case of its predecessor, MWI, MMI must now face the problem of recovering quantum statistics. Obviously, quantum mechanics without the measurement postulate is inherently deterministic, and therefore cannot, by itself, deliver the statistical features we in fact observe in quantum phenomena. Indeed, as far as the core formalism goes, our ignorance regarding the probability of observing any post-measurement outcome could, with impunity, be regarded as equipartitioned to yield equal probabilities for all experiential outcomes. And this is certainly not what we observe in quantum experiments.

To obtain the needed probabilities, MMI proceeds analogously to MWI, which, for the same purpose, postulates the existence of a continuous infinity of worlds.[17] With minds out of reach of the quantum mechanical formalism, however, it does so "nonphysically," as it were, by associating with each brain state in the evolving physical superpositions of measurement outcomes, a continuous infinite set of nonphysical entities or minds. This infinite set is then said to "index" the "chosen" brain state in the sense of somehow relentlessly "tracking" it through a succession of brain states in the evolving composite physical system.[18] The global physical system, of course obeys the Schrödinger evolution, and is therefore superposed. The nonphysical, infinite sets of minds, however, do not obey the formalism and are therefore not superposed. This makes it possible to think of these sets as somehow respectively tracking or "shadowing" the successions of alternative brain states in the evolving composite physical system.

Along the way, individual minds will randomly "choose" membership in one or another of the indexing infinite sets. As the physical system evolves, therefore, there is no telling in which infinite set of minds that "attend" each brain state any one individual mind will end up. The process is therefore genuinely stochastic. Further, the sets involved are postulated as infinite in order to allow the assumption that there are determinable measures over the sets that satisfy the probability calculus and also specify the likelihood, on the occasion of measurement, that any single mind will fall into one or another of the infinite sets that accompany alternative outcomes of the measurement. This then determines the probability that an individual mind will end up perceiving any one of the alternative experimental outcomes.

Finally, this probability can be defined in accordance with the Born statistical algorithm, that is, in terms of the square magnitudes of the coefficients of the component alternatives. The chance that any single mind will end up with any one of the final alternative brain states will correspond to the Born probabilities for each of the possible outcomes of measurement and therefore mirror the correct quantum mechanical statistics. In this way, such a dualistic approach attempts to reconcile the determinate evolution of the superposed physical world with the frequencies we actually observe as definite (discrete) experiences in quantum measurement.

Besides furnishing a venue for securing the desired quantum statistics, positing infinite sets of minds instead of single minds reaps an additional advantage. It avoids a brush with non-local hidden variables.

The reason for this is that single minds can turn out to behave as unwanted, non-local hidden variables in disguise. For, in a single mind theory and under certain conditions—as when two observers are involved at the two ends of a Bell-type experiment—Bell's theorem implies that we must allow strong correlations between the trajectories of the two minds. That is, we must allow non-locality. For, unless we do, the correlations between the two minds will not square with the predictions of QT.[19]

Postulating infinite sets of minds yields still another bonus. It offers a solution to the so-called mindless hulk problem. By offering a superabundance of minds, it avoids the disconcerting possibility in single mind theories that some brains in various branches of the evolving system will end up "uninhabited" by a mind.[20] This possibility, as M. Lockwood points out, would violate the widely accepted principle that minds are supervenient on brain structures. For, there would then be nothing anywhere in the process to differentiate, structurally, the brains that end up with minds from those that do not.[21]

Unfortunately, however, postulating infinite sets of minds, incurs an old risk, namely, non-locality. M. Hemmo and I. Pitowsky have argued that in order to provide a satisfactory solution of the mindless hulk problem, the many minds picture entails the existence of a weak correlation, that is, a weak non-locality, this time, between the infinite sets of minds.[22] Given the unwanted status of even weak non-locality, this objection somewhat threatens the MMI solution of the mindless hulk problem. In the final analysis, however, this version of MMI stumbles most seriously by essentially dispensing with an important relation, namely, that of supervenience. It leaves us with no physical earmarks whatever for telling which individual minds end up tracking which alternative outcomes as more and more superpositions arise in the evolution of Ψ.[23]

It seems, then, that MMI dualism leaves the metaphysical status of nonphysical, "supervenience-free" minds profoundly mysterious and subject to all the puzzles and objections that roil the traditional mind-body issue. And, one might also want to ask at this juncture, does positing infinite sets of nonphysical entities in order to recover quantum statistics make for better theorizing than the standard practice of flatly postulating the Born rule? As D. J. Chalmers points out in a brief critique of MMI, requiring that infinite sets of minds track evolving brain states and prescribing how individual minds evolve is a loss of simplicity. For, implicit in this ontological scenario are ad hoc postulates, i.e., psychophysical and "intrapsychic" laws that have no sort of independent theoretical basis.[24]

An alternative to the Albert and Loewer dualistic version of the MMI is the materialistic one of M. Lockwood and other theorists holding similar views. On this view, the supervenience requirement is dealt with by postulating that the "composite mind" consisting of the many minds associated with a brain is a subsystem of that physical brain. This composite mind is therefore a physical entity that evolves with the rest of the physical system to the very end of the process. Moreover, the evolution is governed by the Schrödinger equation and is therefore deterministic. In contrast to the dualistic version, therefore, there is, on this view, no independent stochastic behavior of minds. Individual minds are *physically* evolving aspects or elements of the many minds "complex" (the "multimind") associated with every sentient being.[25]

From the viewpoint of purely physical theory, such a materialistic conception of mind would seem to be the desirable move. From a more general (e.g., philosophical) viewpoint, however, this sort of slide to materialism entails a world ontology that can seem incomplete and unsatisfying for anyone who believes dualism to be true. It leaves subjective, first-person states (experiences, "qualia") completely out of the picture, thus glossing over very difficult issues posed in cognitive science and in the philosophy of mind by the inexorable fact of consciousness as we know and *feel* its "(what it is like)-ness."

In the face of this omission, Lockwood, throughout, finds it necessary to acknowledge the occurrence of "complete consciousness states" (*maximal experiences*, as he also calls them) that his multimind is capable of "generating." He allows that with any multimind there is associated a continuous infinity of simultaneous experiences of which a single, determinate experience or complete state of consciousness, (maximal experience) is to be associated only with an individual mind.[26] More specifically, Lockwood assumes that there is a set of distinct evolving physical states that constitute a "basis" for the multimind. This basis is a set of physical states in terms of which the state of this "mind" can, at any given time, be expanded—i.e., expanded as a superposition of these states. He calls this basis "the consciousness basis" in line with the assumption that, for each of these basis states, there is a corresponding maximal experience whose subjective identity depends on "what it is like" to have that experience.[27]

The *qualia* of dualistic mind-body talk immediately come to haunt us here and with them the profoundly difficult issue of their status in reality. In final analysis, the materialistic MMI, while, in some foggy sense, allowing the "subsistence" of conscious mental contents or experiences, leaves their metaphysical status deeply mysterious. Indeed, it is never quite clear whether or not experiences are to be thought of as physical entities. If experience is able to "track" consciousness mind bases as they evolve through space-time, then this would make it out to be "somewhat physical."

Lockwood, who painstakingly resists any commitment to dualism, nevertheless seems to suggest that an experience is distinguishable from its quality, namely, the "(what-it-is-like)-ness of having that experience." But then, would this subjective quality be a physical or nonphysical property? As a subjective quality it would have to be a consciously experienced (felt) quality, and again, one would want to ask about the metaphysical status of such an experienced quality. Lockwood tells us that the consciousness basis is "in the (physical) mind" as a composition of physical quantum states each "corresponding" to a determinate experience.[28] The term "corresponding," however, offers no additional clue as to either the nature of the relationship of the quality of an experience to physical quantum states, or the ontological status of such an "experienced quality." What may be fair to say here is that this interpretation construes first-person subjectivity as an essentially "forceless" epiphenomenon utterly incapable of figuring in the physics of the world, yet somehow undeniably present as a "quality" ("the what it is like") or ("what it *feels* like") quality of a physical experience.

As in the case of a dualist MMI, a materialist one must provide the correct quantum statistics. For again, the core Schrödinger formalism is inherently deterministic and yields quantum mechanical splitting or branching but no quantum statistics. Dualistic versions, as we have seen, respond by simply "endowing" free reigning, non physical minds with the ability to "hop on" one or another branch stochastically and to do so in accordance with the Born rule. Recall also that, as in dualistic versions, Lockwood's materialist MMI associates with every conscious being a continuous infinity of simultaneous minds that, as physical entities, differentiate (evolve) quantum mechanically over time.[29] Such continuous infinite sets (this time, of physical minds) associated with every sentient being thus, again, allow for a "natural" measure definable on these sets that both satisfies the probability calculus and, for all outcomes of measurement, recovers the statistics correctly predicted by standard QT.

The minds will undergo their statistically crucial differentiation when the observer looks at the indicator (dial, pointer, or other scenario). At that moment, the proportion of minds that will correspond respectively to the various alternative outcomes of a measurement will mirror the frequencies that standard QT predicts and that we observe experimentally. And again, this is like explaining the observed frequencies of outcomes when drawing from a jar that contains a mixture of differently colored marbles.[30] We assume something about the structure of the mixture, namely, that the proportions along color lines mirror the observed frequencies. As in MWI, each drawing is again a sampling of the multiverse, and the assumptions about "sizes" (measures) of subsets are diagnostic inferences (hypotheses) about its structure.

But though, in this way, probabilities can be tailored as measures of the resulting spaces, one may still ask about the theoretic virtue of the arrangement. In precisely what sense is the vast ontology of minds any more desirable than the adoption of a Born-type rule, *simpliciter*? If interpretations are supposed to inform our grasp of QT, how does positing the "existence" of an infinity of simultaneous minds, nearly all inaccessible, help? As regards physical significance, is such an ontology any less metaphysically burdensome than are the non-local hidden variables it seeks to avoid?

As a further challenge to this account one might refer to the fact that one's personal identity is rooted in a personal biography—a biography comprised of a receding time sequence of past memories, all of them determinate, discrete, unique. Since the consciousness basis in the composite mind consists of simultaneous multiple states, the corresponding multiple experiences are also simultaneous. But, as Lockwood acknowledges, from the point of view of coherent conscious experience, these are compatible (and he must mean this in some phenomenological sense) only if they occur at different times. How then, does consciousness "cut through" the manifold of simultaneous experiences to yield a definite (discrete) biography, to wit, the time sequence of single, complete experiences that we in fact remember? This "tunnel vision," as Lockwood characterizes it, must, it would seem, be left as an inherent limitation of consciousness—an unhinged brute fact of nature—simply there to be acknowledged. In a related context, Lockwood expresses some concerns along these very lines.[31] Once again, as in the case of MWI, one could surely do here with some plausible theory of consciousness for the much-needed details.

Decoherence Theory: Somewhat Promising

Arguably, the two most fundamental and intimately related issues in the foundations of QT may be stated as follows: (1) Much of what the predictions of QT seem to tell us about nature is blatantly counterintuitive and clashes with the most basic standards of intelligibility and coherence in both classical physics and common sense; (2) There persists a seemingly unbridgeable theoretical divide between the quantum mechanical description of the *micro*world and the classical mechanical description of the *macro*world. The first of these issues drives the endless search for a sense-making interpretation of quantum mechanics. The second (of which the measurement problem is a part) fuels certain efforts at closing the macro-micro gap by attempting to provide a reductive explanatory base for the emergence of classical properties from the quantum substrate. This second enterprise, of course, presupposes the universal applicability of QT, and this of course also means the applicability of quantum theoretic terms and concepts across all levels of physical existence.

The two foundational programs are closely entwined. Recovering classical mechanics from QT is only a pipe dream without requisite interlevel principles. These are principles that, if they are to achieve the desired dynamical reduction, must define a causal connection between what we observe on the macrolevel and what we theorize is happening on the microlevel. Such principles must also establish a connection between the language of quantum physics (e.g., quantum state, pure and mixed state, quantum indeterminacy, conjugacy of variables, interfering matter-waves, physically real superposition, expectation value, etc.) and that of classical physics (e.g., classical state, determinacy of observables, definiteness of experience, independence or nonconjugacy of variables, electromagnetic wave, gravitational field, etc.). And all this obviously requires what has not been generally established for QT, namely, a coherent subject matter framework, that is, an *interpretation* or, if you like, "model."

Von Neumann, Bohm, and others unsuccessfully attempted to assume generally acceptable interpretational (ontological) schemes for dealing with measurement and reductive gap closing. A more recent effort along these lines is *Decoherence theory* (DT)—a program that, as might be expected, makes some ontological (interpretational) commitments—but does so without resorting to hidden mechanisms such as collapsing waves, companion waves, or quantum potentials.[32]

DT, however, does not explicitly and systematically propose a new and thoroughgoing interpretation of the Schrödinger formalism in any full sense of "interpretation." What it does, instead, is offer a substantial expansion and "enhancement" of whatever ontological underpinnings mainstream physicists generally assume for QT. In particular, DT takes physical states at all levels (microscopic, mesoscopic, and macroscopic), to be fundamentally quantum mechanical and therefore expressible as superpositions. Moreover, it grants full existential status to all superpositions, that is, all superpositions are assumed to be actual and distinct physical systems. This is sometimes said to be an "existential" interpretation of the superposed quantum states that the formalism delivers. Decoherence theorists, however, acknowledge that how these "physical" superpositions can possibly be observed (experienced) as definite outcomes of measurement remains an open question.[33]

Strongly suggestive of MWI, DT further assumes that such superpositions are, as genuinely physical states, capable of interacting and becoming "quantum mechanically entangled" or "dislocalized. "And this means, as we have noted earlier, that, upon such entanglement, the systems, even after eventual separation, can no longer be described as before, that is, as strictly and separately independent states. In addition and again reminding us of MWI, DT assumes that anyone attempting an observation of such a superposition must become entangled with it, the result being a superposed *composite* system with components that similarly have independent and actual physical existence.[34]

DT therefore brings into QT considerable ontological content with which it hopes to take quantum mechanics beyond kinematics (i.e., description devoid of causal explanation) to the level of a causal dynamics featuring genuinely real physical systems that can interact causally with equally real physical environments. In terms of such a dynamical ontology, it hopes to provide a "solid" explanatory base for confronting the foundational issues.

Entanglement, understood in purely mathematical terms, is implicit in the Schrödinger (core) formalism. For example, consider the distribution of detections at the receiving screen of a standard double slit experiment with electrons. The quantum mechanical calculation for this distribution yields an additional "interference term" indicative of a deep and "wavelike" interrelationship between the electrons and the detecting system. DT interprets this formal interrelationship as an actual physical entanglement resulting from the interaction of two existentially "real" quantum systems—the electrons and the detector. DT holds to the possibility that under certain conditions, relatable to the complexity of the environment, the entanglement of superposed alternatives can result in the "suppression" of interference. It is this suppression that it calls *decoherence.* So, by elaborating a theory of universal entanglement, DT means to work up an explanation of why, in a thoroughly quantum mechanical world, when we observe, we see nothing other than sharply localized, discrete objects. Such an explanation of nature could then provide a link between two dynamically related levels of description, classical and quantum.

Decoherence theorists begin by noting that, as a matter of remarkable fact, the suppression of interference, that is, decoherence, is something that may or may not be observed in nature regardless of level—microscopic, mesoscopic, or macroscopic. Of course, as any student of quantum mechanics knows, the probability distribution of detections at the screen of a standard double slit experiment with electrons cannot, in general, be calculated by classical methods. We cannot assume the electrons to behave classically as entities that, with certain likelihoods, pass through one or another slit. To get the correct answer we must, almost always, calculate quantum mechanically. It is this that yields an interference term presumably indicative of a wave with "two" components that proceed coherently and respectively through the two slits. On the other hand, if we want generally correct answers that are "interference-free" and gotten by purely classical computation, we must detect the electrons at the slits rather than at any place between the slits and the screen.

As a matter of experimental fact, however, interference will, in rare and very special cases, fail to appear even for detections beyond the slits. This can happen if in the region between the slits and the screen an intense environment of quantum systems intervenes. Such a case would seem to indicate that, at any ontological level (macro, meso, or micro), there seem to be physically possible contexts with variables that commute for *all* observables and that therefore will exhibit classical behavior, though of course such behavior will be rare at the micro level.

Correlatively, while the macroscopic and mesoscopic systems are most likely to be interference-suppressed, there seem to be macroscopic systems for which interference is not suppressed. Investigators, for example, have shown the possibility of shielding some superconducting devices from decoherence so as to make it possible to obtain results that strongly suggest simultaneously different (i.e., superpositions of) macroscopic currents. Notably, experimental techniques of the past decade have also made it possible to monitor the suppression of interference for some microscopic and mesoscopic systems. These advances have stirred considerable interest in DT for its bearing on some of the knottiest and most central issues in the foundations of QT.

A very sketchy account of decoherence in terms of the Schrödinger formalism might go as follows: A quantum system, Ψ, evolves according to the Schrödinger equation as a superposition of states. In so doing it may become spontaneously entangled with its environment to form a larger (composite) quantum mechanical system which, in turn, evolves according to the Schrödinger equation.[35]

But as we have just noted, experience suggests, that not all that the formalism delivers is *always* compatible with what is observed in nature. DT theorists therefore assume that within the earlier superposition of states comprising the original system, Ψ, interaction with the environment may induce the "selection" of a subset of "preferred" or *superselected* states.[36] These are states that resist entanglement because they are the states in the superposition that are extra-tightly (narrowly) contained wave packets, (i.e., wave packets of very narrow "spread"). Moreover, these packets can be so narrowly contained in both position and momentum that any measurement that might be performed of the two variables would escape the proscriptions of Heisenberg uncertainty. Such states, therefore, are more stable and particle-like. They are not superposed and are as one might say, less "wave-like" than the others. They are the states for which decoherence has occurred, and this means that for them interference is either lessened or absent altogether.

In this way, entanglement and the resulting superselection leads to the emergence of definite (observable) classicality by "sifting out" all the Hilbert space states except for a classical remainder, namely, a well-defined, and classically structured space with approximately classical properties.[37] Thus, "classicality" is observed in this remainder because the system in point has *spontaneously* interacted with its environment. The resulting entanglement or "coupling" has suppressed interference in the sense that the phase relations necessary for interference have been shifted "upward" to the larger composite (entangled) system, which now consists of earlier state plus environment and which continues to evolve à la Schrödinger. Under appropriate measurements this evolving larger system would be the one to manifest interference effects.

We see then, that the process of measurement is of primary interest in DT, and, as might be expected, is assumed to involve an entangling interaction of the measured system with the measuring device regarded as the "environment" of the measured system. As in the general (non-measurement) case, the measurement interaction induces superselection and results in decoherence—this time said to be "artificially induced." And again under superselection, the result is a "reduced" system consisting of a set of *preferred states* that resist entanglement and for which, therefore, interference is suppressed. Also assumed is the redundant storing of information about these states in the environment. This allows the retrieval of information by the observer when an observation is made. The reduced system is thus further "classicalized" by escaping quantum disturbance and therefore the uncertainty associated with the simultaneous measurement of basic (canonical) conjugate variables. Under actual measurement, therefore, superselection ultimately "displaces" quantum entanglement and yields the stability and definiteness (i.e., the "classicality") that we ordinarily observe in the measurement process.[38]

As we can see, DT theorists draw a very strong analogy between spontaneous decoherence and the "artificial" decoherence in measurement. In the analogy, the spontaneous suppression of interference may be thought of as a generalized sort of "measurement process" in which the natural environment is monitoring or "measuring" the system. Moreover, as in actual measurement, in the spontaneous suppression of interference, it becomes possible to identify some components of a quantum system as "Schrödinger waves" that are very narrowly localized in both position and momentum. With the approximate simultaneous measurement of what are ordinarily noncommuting observables now possible, one can speak of observing a Newtonian trajectory as the wave evolves over time. Examples would be the nearly perfectly Newtonian trajectories of Brownian particles and of alpha particles in a bubble chamber.

So, for DT, the measurement problem becomes generalized as what we may call the *problem of recoverance.* This is the problem of explaining how in the generalized process of "measurement" (spontaneous or artificial) we recover the world of classical dynamics from that of quantum mechanics and do so without invoking extraneous constructions such as wave-collapse, pilot waves, or other devices. The ontological and dynamical constructions that DT resorts to for doing all this, however, remain an area of highly unsettled opinion. In particular, the selection of preferred states (in other contexts, the selection of a preferred basis)—an issue that, as we have seen, has burdened other foundational theory—is one of the most central challenges.[39]

So far, we have discussed DT's theoretical account of a generalized hypothetical process *analogous* to measurement. In the case of an *actual* measurement, DT assumes again an entangling interaction between the systems involved. We are therefore left in the measurement process with a composite system expressed as a superposition of alternative and experientially incompatible outcomes. If the measured system is S, the detecting device D, and the rest of the environment E, this superposition consists of the sum: E entangled with one possible state of S entangled with the corresponding state of D, plus E entangled with an alternative state of S entangled with the corresponding state of D, plus E entangled with still another alternative state of S entangled with the corresponding state of D, and so on.

This, of course, is not the definite (discrete, "singularized") outcome the observer actually experiences. Indeed, the above multiple outcome defies imagining what any experience of it could possibly be like. It is, at least in part, as a response to this measurement puzzle that DT assumes the measurement interaction (with resulting entanglement), to induce within the superposition a "preferred" subset of states. These, as we have already noted, are "robust" states that are not superposed and resist entanglement because they are tightly (narrowly) contained wave packets, that is, wave packets of very narrow spread.

We have here some rather specific assumptions for providing an offhand disposition of the measurement and recoverance problems. But there is nothing in the formalism to suggest the needed dynamical ontology. Interaction and entanglement are indeed well supported by the formalism, but that an interaction should produce some sort of "natural selection" is far from anything the core quantum formalism can even begin to deliver. In no way does it entail any scenario in which coherent states fall by the wayside leaving only decoherent ones waiting in the wings to be experienced as objective states of nature. DT, therefore simply posits the selection process. Such an extra-formal construction, however, is, from the viewpoint of theoretic virtue, little more than an "add-on" the ad hoc flavor of which can be seen to cast some detriment on the DT program.[40]

Even granting DT its superselection, the residue of decoherent states still presents a multiplicity of alternative outcomes. In recognition of this, some investigators have suggested that DT may be understood in the context of an Everret-Wheeler type of MWI wherein all outcomes of measurement are assumed to exist materially as a superposition but perceived separately, in separate worlds as decoherent states.[41] But alas, even when aided with interpretations of this sort, DT would not be relieved of the burdensome postulation of preferred bases.

GRW Theory

In response to the stubborn difficulties that beset the Von Neumann projection postulate, some theorists, a generation or so ago, suggested several presumably more systematic approaches to the measurement problem—approaches that did not give up wave reduction.[42] The idea was to account for the reduction process by introducing stochastic and nonlinear modifications of the Schrödinger formalism. Subsequent refinements of this idea resulted in what is known as the GRW collapse theory (more simply, GRW theory) which has drawn much serious interest.[43]

The authors of this theory start out by noting that the most troublesome superpositions are those that present different spatial locations of macroscopic objects as, for example, two or more simultaneous (and of course incompatible) pointer readings. The idea is that in any objective description of nature, entities (microbodies included) must be uniquely localized ("objectified") meaning by this that they must have more or less well defined positions in space. For assuring this, GRW theorists posit the ongoing occurrence of spontaneous and random processes at the microscopic level of any physical system. These processes are localizations or "hittings," at certain positions, of the elementary constituents of the system. The hittings are governed by a Poisson distribution and occur with an average frequency, f.[44] So far the dynamics are promising.

The trouble starts with the further supposition by GRW that the hittings, in turn, cause strong modifications of the wave function of the system and therefore lead to compelling modifications of the probability density throughout the spaces involved thus creating a demonstrable tendency to suppress linear superpositions of differently localized states.[45] For example, in a superposition containing differing (and therefore incompatible) pointer positions, any spontaneous localization of some constituent microparticle in the pointer localizes the pointer itself (i.e., makes it observationally definite) by a collapsing or reduction—and therefore a suppression—of the embarrassing alternative pointer readings in the superposition. Moreover, since the reduction is a demonstrable consequence of the localizations of constituent microparticles, it is amplified in proportion to the number of hittings and therefore the number of constituent particles in the two alternative pointer reading states that comprise the original superposition.

To complete its account, GRS theory must settle on two key parameters in its formulation. These are the narrowness or accuracy, *d*, of the localizations and their average frequency, *f*. The recommended choices result in a localization every 10^8 years for a microsystem and every 10^{-7} seconds for a macrosystem. This means that while superpositions of microscopic systems evolve with virtually no reductions, superpositions of macroscopic systems—the ones that embarrass the measurement process—occur every micromoment. Finally, the very low frequencies of hittings for microscopic systems and the very large localization width compared to atomic dimensions means that an elementary particle will remain virtually undisturbed by localizations. This helps assure that localizations pose no appreciable violation of the well-tested predictions of orthodox quantum mechanics.

GRW theory represents a brave move away from the flickering "reality" of the Copenhagenists and toward observer-free objectivity in physics. It also means to eschew the "phenomenological incoherences" in the many worlds (or minds) interpretations of QT. Unfortunately, however, though it leads computationally to coherently well-defined states of observable macro-objects (pointers, flashes, etc.), the account remains problematic. Though the notion of hittings has the aura of causal interaction, it is obscured by the recurrent mystery of an interaction between a micro-ontology, which GRW posits as genuinely physical and an abstract wave function whose purported physical significance remains deeply unintelligible. Despite the fact that GRW theorists refer to the Schrödinger equation as "a dynamical equation of motion" and characterize their theory as a "unified dynamics," it remains far from clear, as we have argued throughout, that what the "wave" the Schrödinger equation describes—no matter how it varies algebraically over time—is anything substantive, moving in any kind of real space. Indeed, we have here a metaphysical gap that blocks the desired causal dynamics. The modifications of the probability densities that suppress unwanted components in a superposition so as to recover definite outcomes thus remain a mystery. In final analysis, therefore, QRW leaves QT *algorithmic* rather than *explanatory* in either a strong (causal) or the weaker (unity) sense.[46]

The lack of causal nexus for the collapse process raises still another concern, though a less serious one if we are willing to loosen the notion of causation and thus modify our realism. Recall that the von Neumann wave collapse is a process that occurs simultaneously throughout the full extent of a wave and therefore at superluminal speed. This sort of "non-local" occurrence, for some theorists, raises questions of intelligibility such as those expressed by Einstein in the EPR issue. GRW wave reductions incur the same concerns. They too are instantaneous reductions of purportedly real physical waves and must therefore also be non-local.

Well—it seems time now to ask: Why is there, as yet, no generally accepted *interpretation* of QT, in the fullest (ontological) sense of "interpretation"? And why is settling on one such an irksome and long-standing scientific issue? To approach an answer, let us pause first for some further reflections on the notion of interpretation.

Positivism

There is a minimal sense of "interpretation" formally expressed by twentieth-century positivism according to which a theory is interpreted if some of its non-logical (i.e., descriptive) terms are coordinated with observational subject matter. Such non-logical or descriptive terms are terms such as "particle," "wave," "spin," "location," "momentum," etc., which belong neither to logic nor to pure mathematics. On this view, no theory can have any empirical descriptive content whatever without meaning rules, either explicitly formulated or informally understood, for establishing the required coordination with experience. Indeed, without at least some such minimal "interpretation," QT (or, for that matter, any other theory) would have no physical significance whatever. It would be not physics but, at best, only a formal system of symbols.

In fact, the vocabulary of QT, does, at the very least, suggest reference to existential subject matter such as particles, coordinates, momentum, and spin, though precisely what these terms mean in the context of the theory is fundamentally at issue. This fuzzy "quantum ontology," however, is far too thin to provide any kind of dynamical(causal) structure, though it has been hugely useful, if not utterly indispensable, as a heuristic for the guidance of experimentation and further theoretical development

At any rate "interpretation" at this minimal level is certainly not what is at issue in the present discussion. The interpretation of QT in the positivist's sense is an accomplished fact.[47] Not only does the theory coordinate experiential content with key terms such as "particle," "momentum," and "location"—it also incorporates the prevailing probabilistic interpretation which takes $|\Psi|^2$ to be a probability understood as the relative frequency of some specifiable type of experimental result under a specifiable type of measurement. This establishes QT as an irreducibly statistical theory of quantum systems in which there are no states whose variables are all determinate, that is, "sharp" or dispersion-free.

Interpretation in a Fuller Sense: Ontic Interpretation

A fuller notion of interpretation, as it applies in the present context, comes in two varieties. The first of these is the oldest and most common. It is the idea of incorporating into a theory a physical subject matter or ontology rich enough to provide explanatory content.[48] More explicitly, with the aid of stated semantical rules or, much more usually, just by informal convention, such an ontology is coordinated with the formalism of the theory. The result is an interpretation, constituted of existential subject matter that is somehow assimilable or at least can be linked to our commonsense background. We may refer to interpretations of this sort as *ontic interpretations.*

We need have no qualms here about any vagueness in the verb *coordinate.* In real world physics, the notion of interpretation generally remains more or less implicit. Subject matter is introduced informally either as needed in the elaboration of theory (e.g., positrons in Dirac's theory of the electron) or left undefined explicitly (e.g., mass and time in classical mechanics).

The underlying ontology of ontic interpretations must be fundamental or "basic" in the sense of consisting of systems of elemental, dynamical entities and physical quantities such as time, location, momentum, particles, fields, charges of various kinds, etc., together with their interactive relations. With its aura of familiarity, such an ontological framework is sometimes vaguely referred to as an intuitable or "picturable" or understandable reality model of the theory. This means to indicate that the framework has an "intuitable" spatial structure for expressing causal or other dynamical relations. The *interpreted theory* thus acquires the potential to explain and so qualifies as an *explanatory theory,* that is, a theory that makes it possible to explain the behavior of individuals or aggregates of them either under unifying physical laws of some sort or other or, more strongly, under causal laws.

Ontic interpretations are sometimes said to be "classical" in the sense that the subject matter involved includes the kinds of descriptive attributes (properties, relations) that play out in classical physics. This characterization, however, can be too narrowing and somewhat misleading. The character of the physical "picture" ontic interpretations provide is entirely prior to and free of the distinction between classical and quantum modes of describing nature. The descriptive content in such interpretations generally makes reference to "ordinary" or "commonsense" attributes that are tied epistemically to our basic understanding and turn out to be the attributes we ordinarily and generally believe render a subject matter knowable, intelligible, and real. In this regard, we must note that while virtually all the nontheoretical attributes referred to in classical mechanics are ultimately commonsense attributes, not all commonsense attributes are attributes referred to in classical mechanics.

And now, for interpretation in our second sense, which we may call *formal interpretation.* Such interpretation is the application of some formal scheme—either a logic or an "algebra" (i.e., a mathematical system) for re-expressing the formalism of a theory or for taking it to a higher level of generality so as to enhance its intelligibility or usefulness. Examples are the many and highly controversial approaches to the interpretation of QT, which some authors have referred to as quantum logic.[49] Three-valued logic, many-valued logic, and various types of non-Boolean algebras, that is, algebras that do not obey the laws of standard (Boolean) logic are examples. Formal interpretations, as such, add no new ontological content to a theory and therefore no new explanatory power. They are formal transformations that provide useful reformulations for purposes of explication and possible application.

6

Explanatory and Algorithmic Nomologicals

Of Which Kind is Quantum Theory?

Nomological Types

Well, we might now want to ask: (1) Is there something strategically and conceptually wrongheaded about seeking an acceptable ontology for QT? And (2) Does its ontological deficit indicate—as Einstein once observed[1]—that QT is no more than a mathematical device for prediction, waiting to be replaced by a more enlightening explanatory theory? And finally, (3) Would affirmative answers to (1) and (2) suggest that QT, though remarkably successful for prediction is so conceptually awkward and disunifying from the viewpoint of scientific explanation as to justify a serious and urgent program to replace it with a new explanatory theory? The answers I would propose are "yes" to (1) and (2) but decidedly "no" to (3).

Formal science consists, at least in part, of laws and theories. For convenience we shall refer to these collectively as nomological statements or simply nomologicals among which, for our limited purposes, we can nonexhaustively pick out at least two types—*algorithmic* and *explanatory*.[2] Algorithmic nomologicals are formulations that, by referring to select parts of a system, enable one to calculate some attributes of the system on the basis of others, while expressing little or nothing about causal relations or other unifying content such as patterns and regularities in the subject matter involved. Examples are the laws of Ptolemaic astronomy, the Balmer-Rydberg formulas for the wavelengths of atomic spectra, the Planck quantization rule devised for recovering the experimentally known radiation laws, and the Pauli exclusion principle, as introduced for reconstructing the periodic table of the elements.

Still, algorithmic nomologicals may have considerable theoretical content, meaning by this that some of their terms may make reference to hypothetical subject matter—i.e., subject matter that is not directly observable as, for example, Ptolemaic orbitals, optical rays and wave fronts, matter waves, dimensionless particles, abstract spaces, probability waves, and much more. Depending on how properly embedded their theoretical terms are in some formal structure, algorithmic nomologicals may even rank as theories—i.e., as algorithmic theories.

Our class of algorithmic nomologicals is not a very sharply delimited one, but there are laws which, though lacking causal content, are easily excluded from this category because of their *unifying* content. Examples are geometric optics, the kinematical laws of classical mechanics, Kepler's laws of planetary motion, and the laws of thermodynamics. Due to the unifying patterns and regularities expressed in these nomologicals, they would not, on our taxonomy, be algorithmic.

Algorithmic nomologicals, whether they are laws or theories, can serve splendidly for predicting observational effects, and often they even herald deeper theoretical development. Indeed, these formulations must arise in genuinely physical contexts from which they usually inherit an array of substantive subject matter (planets, orbits, wavelengths, elementary particles, etc.) for ensuring their applicability to nature. But, as already stipulated, such nomologicals express neither causal relations nor much in the way of other unifying content. For saying just how an effect occurs or why it was produced in the first place, we must go to the second of our two types of nomologicals, namely, those that explain.

Explanatory nomologicals make possible what is reasonably taken to lie at the heart of scientific knowledge, namely, *explanation.* To do this, they express the existence of patterns and regularities that give structure of universal scope and intelligibility to subject matter, thus unifying it in the face of apparent diversity and incomprehensiveness. Explanatory nomologicals may also posit existential entities and subject matter such as interacting fields, particles, and spaces of various sorts,[3] all of which, along with "dynamical quantities" such as momentum, position, and spin, make possible an arguably more compelling form of explanation, namely, causal explanation. Indeed, it seems safe to say that *causal explanation* has traditionally been viewed as a "stronger" answer to the question "Why?," than that given by unification alone. In the past half-century, however, the nature of scientific explanation has been a matter of deep controversy reflecting even deeper differences on how to characterize scientific knowledge and truth itself.

Scientific Explanation

The covering law model of explanation, largely advanced by C. G. Hempel in the middle of the last century, took explanation to be deductive subsumption under a covering law.[4]

Accordingly, a fact or a law is considered explained if it can be deduced (derived) from more general premises such as a law or theory, usually in conjunction with a statement about additional specifying (initial) conditions.

The covering law model of explanation, however, immediately drew serious objections.[5] The fact that *Socrates was mortal* is easily deduced from the anthropological law that *all men are mortal* and the specification that *Socrates was a man*. But—as the objection goes—while the law stating the unexceptionable concurrence of the two properties, *human* and *mortal*, and the fact that Socrates is human, do indeed assure us that Socrates is mortal, they do not *explain* his mortality. They do not tell us why his death was unavoidable. This would require the causal laws of biochemistry and physiology about interactions such as oxidation and possibly the destructuralization or random deterioration (aging) of organisms with time. Similarly, one can conclude that someone is pregnant by deducing it from the positive result of a pregnancy test together with a covering biochemical law connecting certain markers in the bloodstream with pregnancy. But it seems quite a stretch to say that the positive test result and the covering law about the test together *explain* the pregnancy, that is, tell us "why" the individual is pregnant. One could on interchanging the terms of the covering law deduce the presence of the markers from the individual's pregnancy. So, as it seems, it is the pregnancy that explains the markers and not the other way around. The covering law model in this case appears to get it backward.

Several attempts were made to sidestep difficulties of this sort by a more general approach in which the unification or systemization produced by covering laws and theories was seen as the essential ingredient in an explanation. Other attempts were made offering pragmatic (contextual), intuitional, and sociological, and even subjective theories of explanation. But many, like myself in 1974 and, later, Wesley Salmon, became increasingly convinced that causation is key to explanation, or at least to explanation in a strong sense—Hume's worries about the necessity of the causal relation notwithstanding.[6] If a prior analysis of causation itself had also to be accomplished, such a program could still be pursued as a distinct part of the philosophy of explanation with full and due attention to Hume's caveats.[7] At any rate, it would seem that the importance of the role of causation at the very core of explanation could not be reasonably denied.

Causation and Explanation

There is much to be said for a causal theory of explanation despite its being controversial to the very present largely due to the stubbornly resistant problem of successfully saying what a causal relation is—a problem that stretches back to Aristotle. The obvious control over the course of events that causal nomologicals make possible needs hardly be mentioned in this context. But the epistemic considerations for placing causation (as generally and perhaps uncritically understood) at the center of explanation are also compelling. Not only do the mechanistic models offered by causal theory seem to deliver an intuitively stronger sense of why certain things happen but the "intuitive grasp" that the objective concreteness of a causal model yields constitutes an enormous heuristic for the growth of knowledge. Theoretical content that can be visualized or otherwise "intuited" dynamically, that is, in terms of actions and interactions, promotes crucial experimentation as well as technological applications, all for testing the theory. But more than this, it can serve the further development and elaboration of the theory itself. Indeed, it is "intuitable," substantive, and typically causal contexts—both theoretic and commonsense—that the history of science so vividly highlights as the fertile birthplaces of scientific creation.[8]

Some of the most historically dramatic examples of this have been the *Gedanken* or thought experiments that have led to monumental breakthroughs at the most fundamental levels of physical theory such as in the discovery of atomic theory, special relativity, John Bell's work, and very much more. In these deeply imagined (some might say highly intuitive) experiments the theoretician arrives at an expected "experimental" outcome, on the basis of background knowledge ranging from commonsense levels to whatever theoretical constructions—often, though not always, of a causal nature—seem applicable. Indeed, though the notion of causation can fade in certain theoretical contexts, explanatory content has always been a crowning feature of any physical theory. And it seems safe to say that, historically, the most general and successful scientific theories from Greek atomism through classical mechanics, physical optics, electrodynamics, and relativity have typically had significant explanatory content.

Felicitously, for our purposes here, we need take no strong position on whether or not scientific explanation is best construed causally. Our ultimate concern is QT which, despite its supreme elegance and spectacular success as an engine of prediction, we shall maintain, fails on explanation in both important senses—namely, the causal and the unificational senses.[9]

Quantum Theory; Born Algorithmic

Surprisingly, QT, historically one of the most elegant and successful of all physical theories, is not an explanatory theory. It was born and remains inherently algorithmic, regardless of the fact that it provides an endless wealth of well-confirmed observational consequences. Its arrival as an astoundingly successful and preeminent part of modern physical theory, therefore, breaks sharply from a long tradition of explanatory science and raises unprecedented issues that have embroiled virtually every one of its major architects.

If we begin historically, we see that the inquiries leading to quantum physics were somewhat unusual. Typically, physical theorizing is a quest for plausible explanations. The reasoning is of the diagnostic sort, and the ultimate aim is the formulation of an explanatory framework that can provide appropriate causal or other strongly unifying relations.[10] This process typically occurs in the context of some puzzle—usually, some set of phenomena that resist explanation in the context of the going theory and of other general knowledge. By drawing on this background—often guided by analogies with known causal relations—and armed with some fresh experimental clues, the theorist attempts to come up with a set of very general nomologicals from which to derive the needed explanatory laws.

Modern physical theorizing ordinarily articulates such a framework mathematically with equations whose symbols designate interacting measurable quantities such as motion, location, vector and scalar fields, forces, charges, potentials, and various space-time variables. These may be conjoined with more abstract content like vector spaces of various dimensions, gauge fields, "curled up" higher dimensions, newly postulated (possibly unobservable) entities, and so on, whose existence is posited in order to deepen, or even first provide the basis for the desired explanatory power. Examples of explanatory theory are Newton's laws of motion, Maxwell's electromagnetic equations, and the equations at various levels of atomic and subatomic theory, relativity (special and general), quantum field theory, and string theory.

QT, on the other hand, was not conceived in quite this manner. There was computational contrivance akin to what sometimes results in algorithmic theory right from the start when Planck tried to find a missing theoretical basis for the known black body radiation laws. The trouble started with a closed hot oven and the infinite quantity that classical calculations yielded for the total radiation. Moreover, classical theory was entirely unable to explain the energy distribution curves at various temperatures—curves that had been solidly established experimentally for "black body" radiation.

Starting with classical electrodynamics, Planck attributed black body radiation, that is, the radiation in an isothermal enclosure, to the absorption and emission of energy by electric oscillators in the walls of the enclosure. He then tried to modify the classical statistical theory for the distribution of energy among the oscillators so that the theory would yield the distribution curves that had by then been well confirmed experimentally. He was, however, unable to do so until he made the bold move that marked the birth of quantum theory. He assumed that the energy of anyone of the electric oscillators must always be a whole number multiple of some fixed quantity. Moreover, this quantity had to be proportional to the frequency, ν, of the oscillator in order to maintain agreement with one of the known radiation laws.[11] This meant that the resulting radiation ultimately had to be quantized, that is, consist of finite chunks (quanta) whose magnitudes at any frequency, ν, were hν, h being the constant of proportionality.

Planck's quantization, born no doubt of much trial and error, was both imaginative and effective. The radiation law it led to worked marvelously for the experimentally known radiation laws. Moreover, it yielded a new law that showed how the waves of high frequency would be sharply suppressed in the higher and higher frequencies and would therefore not contribute to the total energy, thereby limiting this total energy to a finite quantity. The troubling infinity resulting from classical theory and expressed in one of the experimentally incorrect radiation laws was thus made to disappear.

The continuity of energy variation for radiating oscillators, denied by Planck's assumption, was a physical feature imperatively demanded by classical theory. It was therefore something of a revision of the going classical dynamics, and on this basis might be thought to have been a move toward explaining what was then known about black body radiation. As explanatory theory, however, it stood on the slimmest ontological grounds. The positing of discreteness for the energy states of simple harmonic electric oscillators was a computational modification made in a context that was limited to the radiation inside an isothermal enclosure or, at best, also to the oscillators producing the radiation. Meanwhile, the wider ontological context remained largely classical with uneasy consequences for the Planck assumption. Indeed, on the basis of the classical considerations that guided even Planck's thinking, he found his own theory troublesome and revised it to allow oscillators to absorb in a continuous manner leaving only the emission of radiation as a discontinuous process. The "ad hoc-ish" and essentially computational contrivance in these historically monumental developments is quite evident.

The subsequent breakthrough quantizations—four years later, by Einstein and eight years later by Bohr—were ontologically more general and more explanatory in content. However, no overall explanatory quantum theory of matter—i.e., one having a suitable dynamical ontology—was forthcoming for integrating quantum absorption and radiation with these later quantizations as well as with a variety of astounding effects discovered by A. H. Compton, C. Davisson, L. H. Germer, and others. The epistemological gunsights, therefore, had to be lowered, and even the most ambitious theorists were now willing to settle for something less than explanatory coherence, namely, just prediction. The unruly phenomena had to be calculable by whatever formulations one could dream up!

Settling for Prediction

The first formal scheme for doing this was not long in coming. It was matrix mechanics invented by Heisenberg with final contributions by M. Dirac and M. Born. Untypical of mathematical theories in modern science, however, matrix mechanics was blatantly algorithmic. Its inventors proceeded in purely computational fashion by representing the possible energy transitions of the atom as rectangular arrays of numbers (matrices) that serendipitously obeyed the rules of matrix algebra.

Matrices, by virtue of their purely formal characteristics, were eminently suited for representing quantized possible energy transitions. Also, as a matter of algebraic rule, matrices need not commute under multiplication. This means the order in which we multiply two matrices, A and B, can make a difference, so that AB need not equal BA. Accordingly, these theorists were able to represent physical quantities such as position and momentum or time and energy as noncommuting matrices in such a manner that the difference between AB and BA, for any two such matrices, was proportional to a non-zero constant—actually, Planck's constant! All of this strongly suggested that at the quantum level basic physical quantities were to be thought of as noncommuting matrices and no longer, as single numbers.[12] Most remarkably, however, though no ontological sense could be made of this computational (algorithmic) scheme, it worked. It predicted phenomena with astounding and uncanny accuracy.[13]

At about the same time that matrix mechanics was taking shape, W. Pauli introduced his exclusion principle as a means of theoretically generating the known periodic table of the elements. It ruled that no two electrons in an atom can have exactly the same quantum numbers, that is, be in exactly the same quantum state. Again, this add-on rule worked perfectly well for what it was designed to do. Felicitously, in the context of later mathematical developments, it became possible to derive this rule from the symmetry properties of the wave functions assignable to the electron considered as a fermion (i.e., a particle having spin 1/2). But the principle was nonetheless initially adopted as a purely algorithmic device starkly devoid of explanatory content and externally imposed for purely computational purposes.[14]

Only months following the completion of matrix mechanics, Schrödinger, who was bringing to completion what Louis de Broglie had started three years earlier, presented his wave mechanics—a formalism that was no less successful than Heisenberg's, the desired quantizations following from the fact that they made possible the only physically meaningful solutions of the Schrödinger wave equation. The Schrödinger formalism however, was mathematically more interesting, and more physically suggestive.[15] At any rate—thanks to von Neumann's notion of an operator calculus in an abstract Hilbert space—matrix and wave mechanics were readily shown to be entirely equivalent mathematical algorithms.

Unfortunately, however, matter waves, as we have already noted, were ontologically untenable. As a theoretical creation they seemed to have some real potential for prediction (e.g., the possibility of electron diffraction) and for explanation (e.g., atomic stationary states in terms of standing waves). But the lack of bona fide dynamical content (forces, masses, charges, etc.) compromised the physical materiality even of standing waves. What is more, the wave-particle duality they embodied threatened the intuitability—perhaps even the intelligibility—of whatever underlying ontology was already in place. The first move toward wave mechanics, therefore, though starting out with some seeming ontological promise, never quite established a coherent and solid physical basis for the development of an explanatory theory of the microworld.[16]

Indeed, the conceptual genesis of Schrödinger's formulation was decisively algorithmic. Schrödinger began with the momentum of a material particle moving along a path and under a force (due to the particle's potential energy) that depended on its position along that path. The de Broglie hypothesis, relating the variable momentum of the particle to its variable "wave-length," provided him with the notion of wavelength as a function of position. In this way he was able to associate a wavelength with the particle at every point in its path. So far, of course, he was no farther along than de Broglie.

Schrödinger, however, did not stop there. At this point, he broke radically with the classical dynamics of a moving particle and attempted to formulate the means by which de Broglie's "matter wave" could be propagated in space-time as a wave, rather than as a particle in a force field. After all, de Broglie's analogy with electromagnetic waves was shaky at the very start. For though he endowed matter waves with some properties of electromagnetic waves, he could not associate any energy with their intensity. For Schrödinger, then, this may well have been the point of departure from classical materiality. Proceeding in a purely algebraic (algorithmic) manner, he appropriated the partial differential equation for a vibrating string, followed the usual procedure of focusing only on sinusoidal solutions, and further allowed for a wavelength that in general varied with position.[17] The result was a wave equation that provided all the mathematical characteristics he wanted for his matter waves. The final step was simply that of substituting the expression for the de Broglie wavelength into the equation, and the result was the time-independent Schrödinger wave equation $H\Psi = E\Psi$, which featured a *complex* wave function, Ψ, that varied with time and place. If we further introduce the classical Hamiltonian as a differential operator acting on the function Ψ, and, on the basis of certain analogies between momentum and position coordinates, substitute the operator $-h/2\pi i\ \partial/\partial t$ for the energy, we get the time-dependent Schrödinger equation.[18]

Still, though the idea of wave propagation was dynamically suggestive, it left the underlying ontology with a serious gap. There was provision only for an abstract multidimensional space and not for a real physical medium in which to propagate a "complex" rather than real "matter wave." Schrödinger's matter wave therefore remained no less mysterious than de Broglie's.

Nor was the ontological gap filled with Max Born's explicit *probability interpretation*—a strictly mathematical move that came almost immediately after Schrödinger's formulation. Born's interpretation worked impeccably for prediction once translated into expectation values, and the Schrödinger wave thus became a "probability wave" describing probabilities that evolved, with time, in wavelike fashion and in an abstract multidimensional space. The same hard questions facing matter waves, however, remained for probability waves: In what genuinely physical sense could one speak intelligibly of propagating "probability waves" with complex amplitudes in Hilbert space? The whole scheme, however, worked with amazing success as one of the most elegant and powerful prediction engines in all scientific history. But its conceptual development as a computational (algorithmic) device was obvious.

Dirac Transformation Theory

For more completeness of exposition in this general context it seems fair to give some brief attention to a further refinement of the quantum formalism, namely, *Dirac Transformation Theory*.[19] This was a monumental contribution by Paul Dirac that, though adding much *internal* unity and elegance to QT, nevertheless left it no less algorithmic. Dirac came to his transformation theory while thinking very generally about how observables figure in the equations of quantum mechanics. This led him to a theory of noncommuting dynamical variables that interrelated the formulations of Heisenberg, Schrödinger, Pauli, and Born. The result was an incomparably beautiful theory. Combining both matrix and wave mechanical methods, it was conceptually more crisp, more general and more complete than all earlier formulations, thus establishing a new basic unity between them.

But despite its stunning elegance and applicability, Dirac's transformation theory of noncommuting variables was, like its quantum mechanical antecedent, inspired by basically computational considerations—and was thus algorithmic from the very start.[20] What Dirac saw was a formal similarity between the commutator, AB-BA, that comes up in quantum theory as a measure of uncertainty for noncommuting observables, A and B, and the Poisson Bracket that occurs in the Liouville equation describing the evolution in time of a system of particles according to classical statistical mechanics.[21] In order to arrive at a formulation that was mathematically satisfactory and appropriately represented (in terms of the Pauli matrices), the Dirac equation took the form in which eigenfunctions of the energy appeared with an infinite number of negative eigenvalues. This "negative energy," however, was still only a purely mathematical (i.e., formal, computational) result, and a problematic one from the viewpoint of physical significance. The formalism, however, took on some explanatory (ontological) content as soon as the "negative energy" result was interpreted as a new phase of matter, to wit, antimatter.

Spectacular! The Dirac theory of the electron predicted the existence of the positron and, derivatively, of all antiparticles. To express this, Dirac posited a physical model consisting of a sea of negative energy electrons (more dynamical ontology) in which unfilled quantum states or "holes" behaved like positively charged particles. Finally, it was possible to couple Dirac's equation to an electromagnetic field (again, more ontology). On the basis of relativistic considerations and a system of statistics proposed by E. Fermi, the resulting mixture of positive and negative energy components could then be interpreted within the quantum theory of radiation and even allow matter-antimatter creation and annihilation. The theory now had a dynamically explanatory ontology that beckoned strongly in the direction of quantum electrodynamics to which Dirac contributed enormously and which we shall characterize in later discussion (ch. 10) as a "quantumized" fully explanatory theory. (The term *quantumized*, here, is not to be confused with the term "quantized.")

At this point in our discussion it may be still be thought that there was much in both the discovery and development of QT to bring in nomological content with significant unifying, and therefore, explanatory power. What comes to mind in this regard is not only the groundbreaking symmetry considerations by de Broglie that ushered in matter waves but also the theory-building appeal to symmetry in the subsequent elaborations of QT. And aren't appeals to symmetry often moves that typically bring in laws or theories with considerable ontological content and therefore explanatory power?

Symmetry in Scientific Theorizing

In considering an answer, it may be helpful to recall that though the idea of symmetry brings to mind thoughts of unity based on overall constancy and regularities, it is in itself nothing explanatory or even nomological. It is only a very abstract notion that has immensely shaped discovery and practice in the physical sciences. What may have explanatory force is not the concept of symmetry but, rather, some of the nomologicals it can strategically lead to. These are the nomologicals that theorists refer to as *symmetry principles* and that often have the subject matter content needed for explanatory force.

It is hard to think of any guiding strategy, other than the requirement of consistency, that has had a greater effect on the growth of scientific knowledge than the appeal to symmetry of one sort or another. The coherence and comely simplicity of symmetry has lured theorists from the time of Parmenides and the Milesian physicists to the present. And the idea has been to view nature in terms of some deeply unchanging unity or oneness in the face of more superficial change or transformation. In modern physical theory, this "unchangingness" is elegantly expressed as invariance under some set of changes or transformations.

The favoring of symmetry in the art of theorizing, however, is more than the natural preference for the esthetic or for the coherent. It is—one might say—the recognition of symmetry as a signature of "truth" in the sense of its providing major signposts on the way to successful theorizing. Indeed, the role of symmetry in the framing of theory has been crucial since the very dawn of the physical sciences. Throughout the history of physics, symmetry of one sort or another—either as implied by the formalism of a theory or additionally posited for incorporation into that formalism—has played key roles, almost single-handedly, in the creation, elaboration, and even application of theory. In working contexts, symmetry principles can greatly shorten and simplify analysis in solving specific problems. The conservation of energy and momentum laws (these are symmetry principles), for example, facilitate or even make possible the solutions of some mechanical problems—solutions that, if tackled head-on with Newton's equations of motion, would bog down in a thick mire of detail and complexity. In QT, recognizing the symmetries (symmetry *groups*) in the formal representation of a quantum system can help immensely to reduce the hopeless complexity of the Schrödinger equation for many-body systems as, for example, in studying the spectra of complex atoms and nuclei.[22] Further, even in the absence of certain dynamical details, knowing that a system satisfies certain symmetry principles can be useful. By setting up the appropriate representations of the groups that correspond to these symmetries, one may obtain key parameters, that is, some of the quantum numbers and selection rules, for a system, without having to commit to a complete formal analysis.

More basically, however, symmetry principles can provide profound ontological anchorage. After all, in the context of change, coherence calls for something more "basic" to remain steady (invariant)—be it mass-energy, momentum, space-time description, or frames of reference. What is more, the formal simplicity of symmetry can trigger insights that go into monumental theoretic discovery. Historic examples are not lacking: space-time uniformities, along with the great conservation laws—all symmetry principles—mold much of the explanatory, ontological architecture of classical theory. No less notably, the invariance of Maxwell's electromagnetic field equations under a simple algebraic transformation (the Lorentz transformation) is what first cast serious doubt on the physical significance of the notion of a luminiferous ether. Indeed, it led to Michelson's invention of the interferometer and to the famous experiment that followed, with its negative results for a pervasive ether. Relativity came next, and again symmetry played center stage. The equivalence of inertial systems and the invariance of the velocity of light for all observers led to the special theory of relativity and therefore to the requirement of Lorentz invariance for all physical laws. And, following this, the equivalence of gravity and acceleration delivered the keys to the equivalence of all frames of reference whatever their state of motion (general relativity). With the advent of QT and Heisenberg matrix mechanics a more abstract sort of symmetry came into play. This was the celebrated and important symmetry feature of the matrices corresponding to the momentum and coordinates of a quantum system. These matrices have the property of being Hermitian—a property that is key to prediction by bringing the formalism into correspondence with the expectation value of the physical quantity represented by the matrix.

Not much later, de Broglie ushered the notion of a "matter wave" into QT. In a most daring of symmetry leaps, he assumed that, just as there is an electromagnetic wave associated with "electromagnetic particles" (photons), there must be a matter wave associated with material particles.[23] De Broglie's hypothesis could in part have been inspired by W. R. Hamilton's nineteenth-century reformulation of Newton's laws.[24] In geometric optics, one can conceive of a region whose optical index of refraction will vary from point to point, so that a ray of light going through the region will travel with varying velocity. The manner of variation can be made such that a wave packet traveling along the ray will move along some curved path precisely in accordance with the laws of Newtonian mechanics.

Quite apart from whatever inspiration de Broglie may have drawn from Hamilton's formulation, his move was nonetheless a radical "symmetrizing" one. And, like other such strategies, it triggered theory creation—in this case, the formulation of wave mechanics. De Broglie was aware of the simple relationship of wavelength, λ, to momentum, p, for electromagnetic waves, namely, that $p = h/\lambda$. This was a relationship based not only on the Planck-Bohr quantum accounts, but also on electromagnetic theory and special relativity.[25] He then flatly assumed that this relationship, which held for electromagnetic radiation, also held for the matter waves he associated with a particle. In a further move, to some extent anticipating Born's later probability interpretation, he also suggested that the "intensity" of this wave at a given location would yield the probability of finding the particle at that location.

In passing, it is worth noting that de Broglie's assumptions, up to this point in his theorizing, did not take him much beyond the realm of classical description. To bring his account fully into QT, he had to make still other ontological assumptions. First, he assumed that matter waves were small enough to accommodate the Davisson-Germer diffraction of electrons. Then, to explain the Bohr-Sommerfeld quantum condition, he allowed the possibility of standing waves under appropriate boundary conditions.[26] This move towards the quantization of his matter waves was distantly analogous to the quantization of electromagnetic waves by Planck and Einstein.

Following de Broglie's bold assumption, the role of symmetry in QT has continued undiminished. In the absence of relevant dynamical details, for example, knowing that a system satisfies certain symmetry principles can be useful. By setting up the appropriate representations of the "groups" (group theory) that correspond to these symmetries one can obtain some of the quantum numbers and selection rules for a system without committing to a complete formal analysis.

Symmetries of this highly abstract sort have continued to make possible some of the most crucial breakthroughs in the quantum-type theories that followed QT. For example, the imposition of gauge symmetries in quantum field theory (QFT) implies the existence of fields (gauge fields). As quantized entities, these yield the rich ontology of messenger particles (photons, weak gauge bosons, and gluons) for explaining actions of the nongravitational forces (electrical, weak, and strong).[27] More remarkably still, to accommodate the weak force, these field theories, have had to expand the gauge symmetry concept to include a new concept, namely, "symmetry breaking." It is symmetry breaking that leads to a gauge field theory of the weak gluons and therefore of the weak interaction.

Symmetry breaking in turn is implemented by the so-called Higgs particle derived from the hypothetical "scalar" field known as the Higgs field. Field theory postulates this field in order to explain the existence of mass.[28] Without the Higgs field, the theory absurdly predicts that all particles are exactly the same, and without mass. The Higgs field, however, "breaks" some of the symmetry (same mass), thus allowing particles to have *differing* masses. In so doing, this mysterious field acts as a vast repository of energy permeating all of space and interacting with elementary particles as a friction-type force or "drag." The greater the drag, the greater the mass. We have here a rich explanatory ontology generated in the context of abstract (gauge) symmetries—an ontology that, with its diversity of predicted effects, opens up a wide range of possible experiments in particle physics. Finally, in contemporary string theory, symmetry, as we shall see, turns out to be the key to theory development and elaboration. Known as "supersymmetry" this novel kind of symmetry adds a swarm of new particles (sparticles) to the going explanatory ontology. Indeed, it would hardly be an exaggeration to say that it is the appeal to symmetry, more than any other practical strategy for discovery, that has spawned the ontology of virtually all modern field theory.

Abstract Symmetrizing

But despite the hugely important role of symmetry in physical theory (even explanatory physical theory), it would be misguided to view *all* symmetry principles as explanatory nomologicals, that is, as dynamical laws for the explanation of phenomena. Explanatory nomologicals are either causal laws featuring possible interactions in some appropriate space involving such entities as masses, forces, and charges, or they express other existential patterns and regularities. Some symmetry principles, however, though spectacularly effective for solving problems and constructing theory, are not explanatory laws but, instead, express only formal uniformities having no existential descriptive content.

Many of the symmetry moves in modern physical theory have been made on such purely formal and abstract levels. The result has been symmetry principles with neither explanatory power nor even the promise of any. In the elaboration of QT subsequent to the work of de Broglie, for example, the operative symmetry principles, some of which we have already mentioned, have all been highly abstract, algorithmic constructions bereft of ontological content and therefore of explanatory force. The role of symmetry in QT, therefore, though enormously effective algorithmically, has fallen short of imparting explanatory content.

De Broglie's expansive matter-wave hypothesis also fails in a comparable respect. Though on the surface very promising of fresh explanatory content and uniformity, de Broglie's epochal appeal to symmetry was a false dawn ontologically. For it brings in no genuinely existential content. Recall the unresolved issues about the physical significance of matter waves, issues that seriously bring into question their status as actual physical entities. Indeed, on the basis of what is dynamically required of physical waves, matter waves come out as ontologically vacuous and therefore devoid of explanatory potential. The play of symmetry, either as the matter wave hypothesis of de Broglie or as abstract principles of the formalism, therefore, provides no explanatory content, and so, leaves QT no less algorithmic.

Even the notion of spin—introduced in Dirac's new formulation of quantum mechanics in 1928 and operative in the symmetry moves of later quantum dynamics and string theory—does not quite provide QT with any explanatory, ontological moorings. Spin was attributed to dimensionless electrons, and though associated with measurable physical quantities in certain types of experiments, it was "physical spin" only as a heuristic analogy. Any spin as actual rotation that could explain what is observed in these experiments cannot be consistently predicated of a dimensionless electron. The electron therefore behaves only *as if* it had a physically "actual" angular momentum and an associated magnetic moment.

7

A Modest Proposal

Interpretation: A Failed Program

With QT remaining essentially algorithmic throughout its development and with concrete dynamical subject matter virtually unassimilable to the kind of formalism that structures the theory, it is not surprising that no approach to its interpretation has been quite satisfactory. And indeed, nomological algorithms are generally formulated only for predictive purposes with little concern for causal explanation. Typically they deliver little more than the experimental regularities they are designed to predict, and often they are conceived serendipitously on the basis of vague and distant analogies. They may, therefore, not be expected to provide any clues for incorporating the dynamically interacting subject matter required for explanation.

In particular, the core formalism of QT, whether in matrix or wave mechanical form or both, is well characterized in this manner. In all respects, it is neutral, providing no clue whatever to any underlying dynamical mechanism. To wit, it implies nothing about how to approach the problem of measurement, that is, whether to characterize measurement in terms of wave collapse or in terms of a superposed, entangled union of object and apparatus evolving deterministically according to the Schrödinger equation. Again, the crucial deficit here is the total absence of a dynamical ontology but, even more seriously, the highly questionable possibility of bringing one in, uniquely and consistently.[1]

The Schrödinger wave formalism fares no better. It too fails to provide any clue for assigning dynamical content. For, as we have seen, matter waves and probability waves are untenable as dynamical notions. Indeed, as we have noted, the quantum formalisms are algorithmic constructions that not only fail to provide structural clues for any ontic interpretation, but their very nature—be that one of matter waves, probability waves or spins—seems to preclude any assignment of real (ontologically genuine) dynamical content.

The resistive nature of the quantum formalism to ontic interpretation brings to mind more recent attempts at interpretation, this time of the purely formal type. Notable among these is an account by J. Bub in which he quite properly maintains that the quantum formalism is a non-Boolean structure. That is, it does not conform to the Boolean (standard) logic underlying classical theory. What is more, the quantum formalism is strongly non-Boolean in that it cannot be embedded in a non-Boolean structure. Bub also points out that while the Boolean structure of the classical description depicts a world in which actual physical properties evolve in the context of a fixed set of possibilities, the quantum description finds a place for the evolution not only of actuality but also of what is possible and of what is probable.

In a remarkably detailed formal study, Bub shows how a non-Boolean world provides the formal structure for the desired dynamics of possibility and probability—a dynamics in which these two modalities evolve or change with time as tracked determinately by the time-dependent Schrödinger equation.[2] Accordingly, in the world as described by quantum mechanics the set of all possible properties can be represented as some appropriate substructure of the overall non-Boolean property structure. Each such substructure is defined by a quantum state and some definite observable of one's own choosing (a "preferred observable") which is stipulated as determinate. The dynamics of change in a quantum world then consists of tracking the evolution of the quantum state by means of the time-dependent Schrödinger equation. In this way, the temporally evolving quantum state with respect to some preferred observable makes it possible to track the evolution, over time, of the dynamically evolving set of possibilities and the dynamically evolving probabilities that are defined on that set. So, the evolution of what is actual also delivers both the evolution of what is possible and of what is probable under the constraints of the Schrödinger equation.[3] The result of such an analysis is a successful avoidance of the measurement problem and, finally, an observer-independent, no-collapse "interpretation" of quantum mechanics.

Despite the remarkable elegance and beauty of Bub's account, it is subject to several caveats. To begin with, it is a *formal* interpretation, not an *ontic* one, and, as we have already noted, such interpretations do not yield explanatory content. What we have characterized as the algorithmic nature of QT, therefore remains unchanged. Moreover, while such a formal interpretation serves splendidly for delineating the implicit syntactical (formal) features of the theory and doing so with much gain in clarity, it leaves essentially untouched all semantical and ontological issues regarding the kind of reality that the formalism of the theory is supposed to refer to. From the viewpoint of an objective or Einsteinian realism, it therefore gets no closer to dealing with the intelligibility issues regarding that "reality."

And as for the measurement problem, what is accomplished must be similarly qualified. Quite properly, Bub places the measurement problem at the heart of the problem of interpretation, and his account does indeed sidestep the problem. It does so, however, essentially by fiat. The desired definiteness of the outcomes of measurement in the face of superposition is, in the final analysis, the result of stipulation. Determinacy is required in the choice of "preferred variable" so as to recover, for example, the definiteness and stability of a measuring pointer in actual physical measurement. Such definiteness is secured, however, by simply assuming it rather than by any derivation from the theory. It is therefore no less an add-on than von Neumann's picturesque wave collapse, without the latter's intuitive appeal and cachet as a concrete mechanistic model. And there is a further concern. Despite the purportedly observer-free nature of the interpretation, the role of a sentient, choice-making being in determining the outcome of measurement—in other words, the role of consciousness—is not entirely eliminated.

Further concerns remain regarding the metaphysical status of possibility and probability as "evolving modalities." What strange sort of dynamics are the dynamics of possibility and of probability "waves"? The range and magnitudes of "real" possibilities and probabilities can certainly vary with time. This, however, can only be the result of temporal changes in the structure of the existential physical substratum—changes that are governed by time-dependent physical laws. In what sense are possibility and probability waves existentially physical subject matter, so that their propagation over time is genuine physical motion? What are the propelling forces and the displacements in real space? Probability spaces measured on some set of possibilities are informational and not physical subject matter. In today's popular scientific culture we tend to blithely "dematerialize" DNA by calling its structures "information" and we reify information by speaking of its "dynamics." The practice of conflating information with physical reality, however, though fashionable as metaphor, can also lead to philosophically awkward conclusions. The long history of efforts to frame formal interpretations of QT by many distinguished theorists such as J. M. Jauch (my teacher of several decades ago at Palmer lab Princeton) and C. Piron[4] raise very similar concerns and, seem to leave QT no less algorithmic than it has been since the time of its birth.

Well, does all this mean giving up on interpreting QT? An answer of "yes"—and there is much to be said for such an answer—need not put the microworld beyond the reach of a dynamical physics of nature. Surely, the task of making sense of the microworld need not be hampered by construing it exclusively as the problem of interpreting QT. A long century of unsettled and, at times, intensely clashing opinion on the issue of interpretation, would seem to counsel otherwise. Indeed, the history may well be seen as strongly suggesting that cultivating new explanatory theories of nature, rather than attempting to make explanatory sense of QT—at least, in its present state of formulation—may be the more fruitful way to go for the investment of creative effort. Some formalisms, because of their structures (or lack of structure), simply do not allow the possibility for any coordination of intelligible dynamical content, and quantum mechanics is one such formalism.[5] Recent history cautions that ignoring this consideration can be costly. The search for a generally acceptable interpretation of QT, as *cause célèbre* can lose sight of the resistively algorithmic nature of QT and easily resort either to scientifically grotesque constructions or to the neo-Berkeleian subjectivisms and observer dependent ontologies that Einstein along with other objective realists found no peace with.

Away from Interpretation

Turning away from the problem of interpreting QT, however, entails no lapse or reversal of modern trends in basic theory development. It is hard to imagine any physicist not welcoming a fresh and full explanatory theory of microphenomena. We mean here a theory that is sufficiently endowed, both formally and ontologically, to address the mysteries of the physical world with dynamical characterizations of individual systems and processes, rather than with merely a predictive statistics of aggregates.

More specifically, any new theory (H) of microphenomena does not have to be a deepened or extended QT. That is, the formalism of H does not have to implicitly "include"(i.e., entail) any part or aspect of QT so as to render that part or aspect mathematically derivable from H and, in this sense, "explainable" *by* (perhaps better, *within*) H. Such a requirement would only take us back to the burdens of the HVI program and call for what the hidden variable theorists wanted and never quite got. Recall, here, that Bohm tried to relate the quantum state vector $|\Psi>$, representing the state of a quantum system to his posited hidden variable, $|\lambda>$ and to do this, he introduced a nonlinear term into the Schrödinger formalism. In the resulting framework, the result of measurement would be determined by initial values of |Ψ> and |λ>. In a fresh approach to new theory, it would, instead, suffice for H to provide a coherent, causal-mechanistic architecture for explaining quantum phenomena (i.e., the phenomena observed with bubble chambers, colliders, scattering devices, slits, etc.) and all that we already know of the microworld.[6] Certainly, therefore, it would have to recover all of quantum statistics, for they are facts of nature. Would H be a hidden variable theory? *No*, if we mean that H would be logically connected to the quantum formalism in some manner of entailment—but very likely, *yes*, if we mean that it would probably harbor hidden variables of one sort or another.

Hidden Variables Are OK

As regards hidden variables, however, H certainly wouldn't be a first. The history of theoretical science is a history of hidden variables. Atoms, genes, and various familiar particles such as protons, neutrons, electrons, neutrinos, and quantum mechanical spin all, for a time, had to be considered unobservables, and therefore hidden variables, under the available experimental frameworks. Indeed, as important a chunk of explanatory physics as statistical mechanics for deriving the gas laws is a hidden variable theory, and so is optics, which features light rays and other unobservable constructions. In gauge field theory, the fields associated with "unbroken" gauge symmetries are deeply "hidden."[7] Similarly hidden are quarks, which, as hypothetical constituents of observable microparticles, are so tightly interlocked as to be beyond all possible reach of experiment. To all this we can of course, add such *failed* "entities" of the long past as vision rays emanating from the eyes, disease-causing humors, vital forces, phlogiston, caloric, and the venerable luminous cosmic ether.

Our more contemporary theoretical contexts are swarming with still other hidden variables such as gluons, weak gauge bosons, primordial quark-gluon plasmas,[8] gravitons, dark matter, dark energy, and the Higgs field with its associated particle, the Higgs boson.[9] And of course, for each particle type there is also an anti-particle type. String theory contributes generously to this list with such arcana as looped strings, branes, and "tightly curled up" extra dimensions, all together constituting a spatial fabric that "tears" apart and then repairs itself. The incorporation of supersymmetry in present-day superstring theory extends our list still farther with a full compliment of problematic supersymmetric particles (superpartners or "sparticles") such as squarks and selectrons, photinos, gluinos, winos, zinos, gravitinos, Higgsinos, and more![10]

Because of their questionable observability, the significance or meaningfulness of hidden variables has been questioned. But, as some have said, even children understand extrapolations like: "little things so small they cannot be seen." The objection of meaninglessness on the basis of nonobservability, therefore, cannot be reasonably sustained. Indeed, hidden variables can function not only meaningfully but also as the ontic vanguards of scientific discovery in virtually all explanatory theorizing. And, despite the irksome metaphysical auras that the positivistically biased scientific community tends to see in them, they have both instigated and enlightened physical experimentation beyond measure. Without doubt, they have shaped the growth of empirical science as much as any other product of human imagination and creativity, and they have done so even when they have occurred only speculatively in philosophy and in proto-scientific theory such as ancient atomism and early thermodynamics. Of course, every postulated hidden variable is itself a new mystery, a new black box in the ever-receding stream of mystery-piercing black boxes. But this in no way detracts from the power of physical theories fleshed out by hidden variables. It only spells out the fact, well borne out by intellectual history, that, despite any popular claim to the contrary, there cannot, in principle, ever be an end of science.

No End to Science

It may be worth pausing here to note that the remarkable scope and success of contemporary physical theory, together with all that it promises, has prompted some popularizers to proclaim "the end of science," meaning by this that relativity and our quantum-type theories leave nothing more of any basic nature for science to discover.[11] Others, vaguely sensing that there is something wrong with this claim, have tried to buttress their disagreement by laboriously pointing to such unfinished tasks as sequencing genes and codifying their functions, conquering disease, refining cosmological historiography, explaining consciousness, and so on.

Oddly, both parties in this debate skirt past what theorists have generally tended to regard as the very heart of science. This is *explanation*—the process that yields understanding and makes science more than prediction of the future or description of the past or present. Explanation requires a dynamic ontology whose reality elements (e.g., fields, particles, force, mass/energy, space, time, etc.) are customarily postulated in some theoretical context as "givens" or elements of nature. Science, however, takes a raincheck on directly confronting these givens. Sooner or later, often triggered by some puzzling experimental finding, the eternal scientific question can always be asked: "Why these particular givens?." This is a call for "deeper" explanation and therefore for a deeper ontology. Every "deeper ontology," however, leaves room for a still deeper one.

Contemporary string theory, for example, promises to explain all the givens of the standard model (i.e., the existence of all the subatomic particles and their properties). It means to do this in terms of the various modes of vibration of elemental strings and their higher dimensional derivatives. The theory has therefore been called "the theory of everything" (TOE). The understandable exuberance in this characterization, however, overlooks much. String theory explains (recovers) neither the rules of QT nor its predictive efficacy. Nor, again, does it tell us *why* some strings vibrate in one mode and others in another, or why the dynamical and space-time parameters that determine vibrational modes have *those* particular values (or configurations) and not others? And we can still ask: Why are there string loops in the first place? Or, are strings themselves composite? And so on. Indeed, string theory cannot be the theory of everything; certainly, it cannot explain itself.

8

Quantumization

The Quantum Supplementation of Explanatory Theory

The Quantum-Theoretic Context

Now, where does all this leave QT, whose mathematical depth and elegance are without precedent among algorithmic theories and whose predictive power has made it a major player in all modern physics, despite its mind-numbing puzzles and explanatory insufficiencies? As the only encompassing predictive algorithm known for microphenomena, QT is certainly here to stay, probably for a long, long while.

Indeed, physicists have clung to QT in the formulation of what we may call "quantumized" (not to be confused with "quantized") explanatory theory. The formulation of quantum field theories over the past several decades and, more recently, the development of string theory illustrate a vastly successful amalgamating strategy for the structuring of physical theory. This is the practice of developing new explanatory theory in the full context of quantum mechanics, that is, by simply "usurping" the rules of QT. This practice can be a vastly empowering one by either permitting or excluding, in purely computational (algorithmic) fashion, various types of properties, relations, or processes. An example is the introduction of quantum mechanical spin for the point-particles of the standard model in quantum field theory and the appeal to the uncertainty principle to make possible the spontaneous eruption, at sub-microdistances, of dynamically active virtual particles that survive on borrowed energy for sub-micromoments—all of which, as we well know, are utterly impossible allowances in non-quantum contexts.[1] Other examples are the symmetries (supersymmetries) based on quantum-theoretic properties—symmetries that are clearly of a quantum mechanical sort. These have been crucial in the elaboration of both QFT and string theory.

In the quantum supplementation of physical theory, theorists generally presuppose a quantum mechanical context (i.e., the rules of QT), any part of which they can use as needed in conjunction with theories framed in ontologically concrete dynamical terms. This, of course, means having to live with the gnawing perplexities of quantum phenomena and QT, which this strategy leaves unresolved. Non-locality, for example, would have to be unexplained, but allowed, for systems that have interacted in certain ways or have been prepared so as to become quantum mechanically entangled or "coupled." Entanglement and the uncanniness of non-local influences that result from it do challenge our commonsense intuitions and therefore seem somewhat "magical." The challenge, however, is not insuperable. Non-local "influences" are still analogous to causes and can therefore be said to have a measure of scrutability. In the context of a liberalized, causally tolerant ontology, therefore, our deeply ingrained, commonsense or classical realism can give way to a "realism" that allows more intimate correlations within its reality.[2]

The challenges of the uncertainty principle, another major item in the quantum patrimony for quantumized theories, are still harsher by clashing with the classical and commonsense notion of what it is to be a definite particle with a definite trajectory. But again, under a modified realism, a calculable, though uneliminable, fuzziness at quantum levels is not entirely unfathomable, especially in view of all the supporting evidence for it. Similar "intuition-allowances" can be made for other quantum features such as quantum spin, the creation of virtual pairs on borrowed energy, the lumpiness (quantization) of energy, and the probabilities governing the phenomena involved. While these features may be considerably unsettling from a classical viewpoint, theorists can and do nevertheless incorporate them consistently in explanatory theories like quantum field theory and string theory. But once again, any philosophical peace with the resulting theories requires the objective realism that strictly underlies classical physics to yield to a liberalized or critical realism that can live with the desired quantum features and allowances. Happily, the practicality in such "quantum tolerance" can make retreating from the arguably futile quest for a generally acceptable interpretation of QT philosophically more palatable.

Quantum Electrodynamics

The quantum supplementation of explanatory theory is more than a remote possibility. It has been superbly achieved in a highly successful example of more than fifty years standing, namely, quantum electrodynamics (QED), whose predictions have been successful and precise to an astonishing degree.[3] In formulating this theory, the aim was to incorporate the laws of special relativity into QT and, in the context of such a relativistic QT, provide an explanatory dynamics for the interactions of electromagnetic radiation and matter. This was done by utilizing the ontologically rich and essentially classical notion of field. QED deals with "free" and interacting electromagnetic (Maxwellian) fields and fields of a special type conceived by Dirac and therefore called "Dirac fields."[4] In this case, however, these fields are quantized. The particles resulting from the "particleized" electromagnetic field are identified as photons. Such photons are conceived as "messenger particles" that transmit or, better, *mediate* the force that governs the interactive behavior of electrically charged particles and fields.[5] The uncertainty principle and quantum probability are necessary for correctly predicting that this force of unlimited range diminishes with distance. Special relativity, on the other hand, provides for the interrelations between the momentum, energy, and motion of the messenger particles.[6]

It seems appropriate to mention here that though remarkably successful, QED did encounter what is by now a historically notorious internal difficulty. In calculating values for the properties of the electron, troublesome singularities (infinite values) appeared that, of course, could not be given physical significance. A purely algorithmic remedy known as *renormalization*, however, was invented for suppressing these singularities, thus "rescuing" the theory. The strategy involved was to redefine certain physical constants and do so consistently. This worked, but only in a qualified sense. For despite its great Band-Aid utility, renormalization was nonetheless a purely formal, fiat-type, ad hoc means of papering over what some theoreticians saw as a flaw somewhere in the theory.[7]

Quantum Field Theory

A more sweeping example of a quantum-type of explanatory theory is quantum field theory (QFT) which encompasses all three fundamental, nongravitational forces—electromagnetic, weak, and strong. It was conceived by applying methods similar to those of QED. This time, however, new types of symmetries known as electromagnetic, weak, and strong gauge symmetries are required of (imposed upon) the electromagnetic, weak, and strong interactions, respectively.[8] Gauge symmetries are abstract formal symmetries (not of a space-time sort) that render the interactions between particles invariant to certain shifts in their force charges. Maintaining the invariance requires compensatory effects and this requirement is remarkably satisfied if the existence of the three nongravitational forces is assumed![9]

Deep theoretical analysis further reveals that the three nongravitational forces, which at familiar distances differ so widely, tend to converge to a common strength as we approach sub-microdistances such as 10^{-31} meters. But the convergence is not quite complete, and the discrepancy, though very small, is theoretically unacceptable. As before, however, symmetry—of a new and again abstract kind called *supersymmetry*—comes to the rescue. Novel about supersymmetry is the fact that it is based on the quantum theoretic property of "spin"—a property that is classically meaningless for point-particles, that is, particles without dimensions. Still, experiments with atomic spectra have long suggested and QT allows that such elementary particles, though classically incapable of rotation, have something very much like spin, "geometrically abstract" though the concept may be. This somewhat abstract "spin," in turn, makes possible a symmetry that, like other gauge symmetries, can be expressed only abstractly in terms of higher dimensional spaces.

The role of supersymmetry in QFT is theoretically decisive. Besides expanding the ontology of quantum field theory extensively, the incorporation of supersymmetry entirely eliminates the convergence discrepancy in the forces when they are extrapolated to Planck-level microdistances. All three nongravitational forces—electrical, weak, and strong—are thus brought to convergence under one unifying explanatory framework. (This dramatically highlights the power of supersymmetry and foreshadows the remarkably crucial role it later plays in the elaboration of string theory.)

The full development of QFT yields the *standard model of particle physics,* which, within a quantum mechanical framework, includes three families of basic point-particles (fermions) variously interacting under four basic forces that are mediated by transient force-carrying "messenger" particles (bosons).[10] This rich ontology (which also includes antiparticles for each of the fundamental particles) forms the dynamical base of what is by now a highly successful explanatory theory.[11]

The standard model, however, with its point-particle concept, fails to incorporate the fourth fundamental force, *gravity*. Indeed at ultra-microdistances (sub-Planck levels), the description of nature given by QFT clashes with that given by general relativity, which models space-time deterministically and smoothly (continuously) at all space-time levels. By contrast with general relativity, the quantum aspects of QFT are probabilistic and, at sub-Planck levels, they model space-time as turbulent and discontinuous. Moreover, as a "quantumized" theory, QFT incorporates Heisenberg uncertainty. This means that, for the very short time intervals (Δt) that elapse over very short distances, there is the expectation of large and abrupt energy (ΔE) fluctuations, 't' and 'E' being conjugate variables. Indeed, at sub-microdistances, that is, distances shorter than the Planck distance (10^{-35} meters) over which changes can occur in ultrashort time intervals, the changes can be very abrupt. This, then, can result in large enough energy "clumps" to make possible the production of virtual particles, that is, particles that survive only for a very short time (of the order of 10^{-43} seconds).[12]

This micro-mishmash of discontinuities or "quantum foam" as some have called it, though occurring only below the sub-Planck level, nevertheless conflicts fundamentally with general relativity whose equations require *continuous* space-time curvatures for all distances, however small. QFT, therefore, with its quantized fields and point-particles fails flatly to incorporate the gravitational force as described by general relativity.

QFT also fails from the viewpoint of a desirably greater purview and stronger unification. Though the standard model certainly provides a rich explanatory ontology, it leaves the existence of a wide diversity of physical entities as brute facts of microreality. The standard model itself, therefore, cries out for explanation. The grand tasks of providing this explanation and of unifying all four forces is left to the possibility of a far deeper theory, namely, string theory—the most ambitious and sweeping example to date of an explanatory quantum-type ("quantumized") theory.

String Theory

String theory is unsparingly ontological, and does not shy away from a deeply mechanistic, explanatory ontology.[13] It departs, however, from the idealized point-particles of QFT by positing, in their place, infinitely thin but finitely long (one-dimensional) loops whose tiny length is of the order of magnitude of the Planck length (i.e., 10^{-35}meters).[14] As extended (one-dimensional) entities, all strings are structurally identical but vibrate with differing resonant patterns (frequency and amplitude), which, if the theory is correct, completely determine the known properties of the various elementary particles of the standard model. Accordingly, therefore, the physical parameters of any elementary particle—be it a mass particle or a force-transmitting particle—should be deducible from the way its constituent string vibrates.[15] (Notably, the theory, by denying strings their thickness, leaves them without "inherent" mass. This makes it possible for them—when in the appropriate vibration state—to account for the existence of even massless particles such as the photon and the graviton.)[16]

Of course, no purely formal framework can by itself yield (predict, explain) any matter of natural fact. In order to deliver on its promises, therefore, string theory must not only surmount stubbornly resistive mathematical difficulties, it must also be conjoined with some numerically expressed physical quantity (or quantities) that only nature can provide. This, it turns out, is the tension on the elemental strings. Some calculations based on what we know about the gravitational force indicate that this tension must be enormous (of the order of 10^{42} pounds). Its determination and what it implies about fundamental particle masses in nature is a further challenge in the development and possible confirmation of the theory.

The causal dependence of particle properties on the vibrational modes that strings can take on applies to all such properties, that is, mass and force "charges" together with the properties of all messenger particles such as gravitons, weak gauge bosons, and gluons.[17] This means that if we can determine the resonant modes in which the theory permits strings to vibrate, we can deduce (predict, explain) the standard particle properties that heretofore could be determined only by measurement, as the "givens" of nature.

This explanatory reach promises to encompass the entirety of nature, since all phenomena are seen as the result of dynamical string interactions. The strings of string theory may therefore be characterized as "physical strands" or simply "reality strands" posited as ontologically constitutive entities. Unlike "matter waves," therefore, strings and their motions are to be construed as genuinely causal factors and not as algorithmic constructs. The theory offers them as substantial physical entities whose law-like behavior provides a possible basis for the causal explanation of the phenomena we observe.

To avoid certain mathematically absurd consequences, string theory, it seems, cannot remain free of its own "exotica," and this of course can cast a shadow on its plausibility. The theory must be framed in an extended dimensional space-time that allows strings ten degrees of freedom. This then means ten spatial dimensions and a temporal one, thus making for a space of eleven dimensions in all.[18] The difficulties in this framework are obvious. Not only is it geometrically hulky, but it flies in the face of the phenomenological fact that all sense data are confined to three spatial and one temporal dimension. The extra spatial dimensions must therefore be construed as "curled up" into a size small enough to render them inherently unobservable.[19] The theory, to the extent that it is known so far, however, implies certain requirements on how the extra dimensions are curled up. A class of complex topological structures known as Calabi-Yau spaces satisfies these requirements. Some one or other of these is therefore a candidate for the tightly crimped dimensions. When the proper Calabi-Yau space is determined, the theory must then describe physical processes in terms of interacting and intersecting string loops that vibrate with various resonant patterns in that particular space.[20]

To achieve full theoretical development, however, string theory incorporates a quantum mechanical framework. Starting with its novel ontology and with the basically classical equations of vibrating string loops, string theory is thus, as we might say, "quantumized" by incorporating the familiar restrictions, allowances, quantizations, etc. of QT. Accordingly, string vibrations can occur only in discrete units with the possible energy states of a given vibrating string restricted to a whole-number multiple of some minimal amount. The string's tension determines the minimal energy; the amplitude of the vibration determines the basic whole-number multiple.

Hypothetical calculations involving the graviton's vibrational pattern indicate that the force exerted by such a pattern is inversely proportional to the tension of the vibrating string. Since this force is extremely weak, however, the tension must be enormously great—amounting to as much as *thirty-nine* orders of magnitude greater than any tension ever encountered in ordinary experimental contexts. Extrapolating further, this is seen to mean that the minimal vibrational energy (and therefore the minimal equivalent mass) of a string—a quantity that is proportional to the tension—must be far too large compared to anything encountered in the experimental contexts of particle physics. The quantum mechanical framework in whose context string theory has been developed seems to provide a remedy. It allows the production of "negative energy" in the quantum foam at the microlevels (Planck levels) of strings for the cancellation of the vibrational energy to bring the net energy down to the level of the small masses of elementary particles.

The quantum mechanical context in which string theory is developed also provides features such as uncertainty, exclusion, virtual particle creation, quantum mechanical spin, and even the possibility of non-local influences (non-locality). We have here a theory that starts out with the rather simple notion of a distinct, one-dimensional vibrating string, but then admits quantum-theoretic processes such as the creation of virtual string pairs, quantization, and duality symmetry, all of which not only bring in the vastly predictive power of QT but also provide powerful ontological flexibilities and allowances that make for further theoretical development.

The mathematical articulation of string theory and its elaboration, however, are hugely challenging tasks. The basic equations are formulated only approximately, and, to boot, the calculations for extracting their implications are themselves also based on approximative (perturbational) methods. Unfortunately, however, none of this suffices for determining which one of the infinitely many possible Calabi-Yau shapes and sizes characterizes the extra dimensions of string theory.[21]

The mathematical impasses, however, do not end here. For, even if the correct Calabi-Yau space were known, finding some exact solution of the string theory equations would still not yet be feasible. It would require mathematical techniques far beyond anything like present-day levels. And, finding such a solution would be necessary in order to determine, exactly, the frequency spread of string vibrations. Only then, on the basis of *vibrational spectra*, could specific experimental consequences be deduced for deciding whether or not string theory squares with nature. What is more, an exact solution is necessary for uncovering the full explanatory content of string theory. This is the content that would tell us if strings—with their various resonant patterns, interacting and intersecting in some uniquely determined Calabi-Yau space—can, as promised, explain the measured constants of our universe, the known properties of the particles that comprise the standard model, and the ever-accumulating facts of cosmology and particle physics. Even on the level of merely formal construction and elaboration, therefore, string theory has a very, very long way to go.[22]

Despite the absence of the desired solutions, however, a most remarkable and hopeful result is nevertheless available. Theoretical analysis shows that for all possible Calabi-Yau spaces, one particular resonant pattern will always be there. It is the one that corresponds to zero mass and spin-2 and therefore to nothing other than the quantum gravitational messenger particle, the *graviton*. This points to a deep theoretical unity between gravity and the other three fundamental forces, albeit the unity occurs only at sub-microlevels and in the higher dimensions—both beyond all presently possible experience. Indeed, theoretical hints about the closed shape of the vibrating graviton loop also promise an even closer unification with the other three forces by possibly explaining why gravity appears to be so singularly weak compared to the nongravitational forces.[23]

More notably, however, string theory, as it presently stands, does yield at least one qualitative but experimentally promising result. As we shall see presently, the theory cannot be successfully elaborated without the incorporation of supersymmetry. Supersymmetry, in turn, presupposes (implies) the existence of superpartners—particles that may turn out to be observable provided their masses are not too large for the energies of present-day colliders.[24]

The need for supersymmetry in string theory can be seen to arise on the basis of the following considerations. String theory is a quantumized theory. At the sub-microlevel of any basic string interaction, therefore, the uncertainty principle of QT "kicks in" to make possible—with probability—the creation of a transient (virtual) string-antistring pair. (This of course is something categorically disallowed in strictly classical contexts.)[25] An ultra-micromoment later, the pair are likely to merge, then likely to split again, then, in the turmoil of quantum uncertainty, rejoin, then split again, etc. The process may repeat indefinitely to form a sequence of "loops" consisting of joined and disjoined string pairs.

Theoretical analysis reveals that the likelihood of producing any one of these loops is determined by a parameter known as the string coupling constant—a quantity specific for each of several possible string theories and which therefore has much to do with the physics of that theory. The coupling constants of string theories, however, are hypothetical quantities whose values are not presently known. Indeed, whatever physicists do know about the content of any string theory has been determined not by exact methods, but by assuming a coupling constant for that theory and then applying approximation (perturbation) methods. Unfortunately these work only if one assumes that the coupling constant is less than one, or even better, much less than one, in which case the probability of any perturbing loop formations is very small.

For coupling constants not less than one, the likelihood of loop formation is higher. And though constituted of only transient entities, the resulting, ever-extending loop sequence can so perturb the original interaction as to quickly take the process beyond the reach of mathematical analysis. As a result, for coupling constants not less than one, all known perturbation methods become entirely unavailing.

Fortunately quantumization, by allowing spin, comes to the rescue as it did in the case of QFT.[26] Now tailored for strings, the quantum-theoretic property of spin makes it possible to articulate and incorporate the all-important principle of *supersymmetry*—this time "for strings." As a string-theoretic principle supersymmetry relates elementary entities in terms of their quantum mechanical spins and requires that every elementary particle of nature come paired with a superpartner having a spin differing from its partner's by one-half. Because there are several ways of incorporating supersymmetry, five different forms of string theory are possible, each having its own characteristic (but unknown) coupling constant.

We have here an ontology expansion—made possible by quantumization—that is crucial for string theory. It entails a *minimality constraint* that leads to a breakthrough discovery in the elaboration of the theory. More specifically, what now becomes possible is the exact rather than the perturbative (approximate) calculation of a limited (very special) set of physical states, known as BPS states.[27] These are states that satisfy the so-called *minimality constraint.*[28] Moreover, supersymmetry does this for any of the five string theories and for any assumed coupling constant, however strong (large). The properties of BPS states (masses and charges) for theories with large coupling constants, therefore, can be determined exactly despite their inaccessibility by perturbative methods.

This important discovery leads to a remarkable result. It was found that, for a particular one of the five string theories, the properties (masses and charges) of its strong coupling BPS states agree exactly with the properties of the weak (i.e., less than one) coupling BPS states of another particular one of the five theories—and vice versa. This strongly suggests that, in general, the entire physics (not just the properties of BPS states) of these two theories relate in the same way. The strong coupling physics of one of the theories is the same as the weak coupling physics of the other. Or, as we might say, the two theories are "duals" to one another, so that hypothetically increasing the coupling constant of one will transmute it into the other—its so-called *dual.*[29]

This duality, together with another in which one of the string theories turns out to be a dual to itself, is a crucial string theoretic feature that can vastly enhance the theorist's power to elaborate any string theory. Because there is presently no known way of determining the actual coupling constant of a string theory, physicists, in studying that theory, must proceed hypothetically. They assume one or another value of the coupling constant and then attempt to extract some of the physics of the theory, for that assumed value of the coupling constant. In attempting to deduce the physics of a string theory, S, however, perturbative methods (the only ones available) do not work if the assumed coupling constant of S is strong (i.e., greater than one). The dual of S, however, if it can be found, will yield to perturbative analysis, thereby making it possible to fetch some of the physics of the strongly coupled S.[30]

Establishing duality relations between string theories is apparently the key to unlocking the physics of any string theory, at least as far as perturbative methods will take one. Fortunately, the coupling constant dualities that we have noted, together with another class of dualities that arise on the basis of the geometrical form of space-time, results in a network of determinable dualities joining all string theories. These interrelations are strongly suggestive of a basic unity between the five superficially different string theories, which, in final analysis, turn out to be duals not only of one another but also of one core theory—a theory whose elaboration as a *unified* string theory (M-theory) is now a major preoccupation of string theorists.

It is perhaps worth noting, at this point, that the strings of string theory are hidden variables par excellence. Their length, of the order of the Planck length or 10^{-35} m (i.e., 10^{-24} times the size of an atom), is far too small by enormously many orders of magnitude to be detected by any divinable future method.[31] Still, they and their higher dimensional (also "unobservable") derivatives (branes) vibrating in a realm of extended dimensions, constitute a dynamical ontology whose "logical shadows" (i.e., implications or predictions) are nonetheless about the phenomenal world of experiment and observation. The future detection of some superpartner, for example, could constitute a good measure of evidence for the existence of strings.

Unremarkably, string theory, like many other theories is a bona fide hidden variable theory. What it is not, however, is a hidden variable interpretation of QT, designed to make dynamical sense of the puzzling aspects of quantum-theoretic predictions. As a quantumized physical theory, that is, one that is formulated in the context of QT, the predictions of string theory must be consistent with those of QT, which, indeed, are facts of nature. No string-theoretic computation, however, can yield any part or aspect of QT's formalism in some reductive or other (deductive) explanatory sense. String theory incorporates QT simply as a conjoined supplementation and is therefore not a Bohmian hidden variable theory.

With its hidden variable ontology and its quantum mechanical rules, string theory presents a highly ambitious program. It aims to go beyond the unification of the fundamental, nongravitational forces (electromagnetic, strong, and weak) as achieved by QFT. This means it also aims to encompass, as well, the fourth fundamental force namely, gravitation, and, with that, general relativity within which gravitation is so deeply embedded. We have already noted how the persistence of the resonant pattern corresponding to the bosonic graviton (spin 2) throughout all possible Calabi-Yau spaces hints at a deep theoretical unity for the four fundamental forces. What is further required for this unity, however, is remedying the near-century-old conflict between general relativity and QT. The choppy discontinuities and singularities resulting from Heisenberg uncertainty at sub-Planck-length distances violate general relativity, which requires continuity for all scales however small.

Now, when the messenger particles in an interaction are gravitons, infinities arise for vanishingly small separations between interacting particles. On the standard model, however, all elementary particles are dimensionless points. The magnitude of their separations therefore can unallowably vanish. The fact, however, that strings are irreducibly extended entities—a radical departure from the point-particle ontology of QFT—provides a resolution of this difficulty on the basis of both relativistic and ontological considerations. The string length makes all elementary interactions subject to simultaneity-at-a-distance determinations. And though in this case, the lengths are ultra-microscopic, they are nevertheless finite and therefore define a real separation of points on the interacting string loops. The place-time of interaction is therefore different for different observers depending on their motion relative to each other.

This relativity in the place-time of any string interaction raises the ontological or, if you will, "existential" question of where, in space-time, the string interaction actually occurs. The notion involved here is the elusive one of simultaneity at a distance. On the face of it, the idea of two spatially separated events happening at the same time seems clear enough. The trouble begins when we try to say how any observer would determine the simultaneity of two such events. This requires defining simultaneity operationally in terms of light paths and midpoints of separation. Reflection easily convinces that for two observers in frames of reference that are in motion relative to each other, the determinations of the simultaneity must differ. This means that in the full context of our space-time and other physical concepts, it is impossible to say what sort of experimental operations could ever establish simultaneity at a distance as an absolute determination for all observers, independently of their relative motion. The ontological move occurs at this point. As regards what value to put into the equations for the time of the string interaction, the theorist decides that, in fact, there is no such "single" quantity holding for all observers, regardless of their motions. The desired quantity is therefore seen as ontologically indefinite and operationally unavailable as a well-defined quantity for use in string dynamics. This "operationalization" of the time-place of string interactions thus algebraically eliminates or "smoothes out" the unruly singularities, thereby yielding a mathematically well-behaved, finitized, and relativity friendly outcome.

We have, then, two results: (1) The properties of the bosonic graviton (massless, with spin 2) emerge as the very properties of one of the resonant patterns of a string loop; (2) The relativistic considerations arising from the circumstance that strings have extension eliminate the clash at sub-microlevels between QT and general relativity. The force of gravity, which resists incorporation into QFT, is thus seen to come out consistently as a natural part of the M-theory ontology.[32]

More sweeping still, M-theory, as a quantumized theory, aspires to be "a theory of everything." That is, the various vibrational patterns of string loops and their derivatives will, hopefully, yield the values of nature's "given" constants. And this would include the masses of all elementary particles (baryonic and non-baryonic), the permeability and permissivity of free space, the gravitational and fine structure constants, and more—perhaps even experimentally elusive quantities such as the cosmological constant. The great overall promise, then, is a theory that not only predicts but also *explains*! The input factual content needed to accomplish this would be data leading to a knowledge of the tension in string loops. This would, in turn, help determine the frequencies of vibration and a minimal (quantal) string energy.

In reflecting on the gargantuan sweep of string theory it is important to keep in mind that the project is still largely only a promise. The working results, so far available, do not point toward anything like a unique description of the physical microworld. Indeed, the number of possibilities in this regard—each describing a different possible universe—remains staggering. The core ontology of string theory—loops, membranes (branes), and dynamically operative extradimensional spaces leaves unspecified many possibilities. Loops may have many sizes and configurations in their multidimensional environment and can vary in a limitless number of ways. Add to this the fact that the Calabi-Yau shape of the extra dimensions may have any one of an infinite number of configurations and the range of possibilities, despite the fact that the resonance modes are quantized, becomes limitless. In order to narrow down this range, therefore, more content specification, perhaps based on astronomical observations, may have to be infused into the theory, via, possibly, new formulational techniques.[33]

The attempt to deal with any of this has generated a rather murky speculation that has twiddled the imaginations of some otherwise serious theorists. It is the so-called "Cosmic Anthropic Principle" one version of which seems to say that, whatever may be the implications (solutions of the equations) of string theory, they must at least lead to laws that allow life to exist. This claim, though true is trivial. Obviously, the laws that theorists are attempting to extract from string theory must be permissive of life and certainly of human life or else they would be false. (The theorists themselves, as live beings, would not be there to make the attempt except possibly by "miraculous dispensation." Indeed, the "principle" could be made even stronger (though no less trivial) by adding that string laws must allow the emergence of consciousness and *intelligent* life.

From the viewpoint of theory construction, however, the requirement that the nomological content of string theory be life tolerant seems too general to be of much help, though it does select for a limited class of possible dynamical systems or "universes." One can easily grant that the existence of life is sensitively dependent on the values of some of the known constants of nature. Yes, every indication has it that the range of variation of these constants, within which anything like biological phenomena can occur, is very, very narrow indeed.[34] But the number of remaining possibilities ("possible universes"), even within this narrow band, seems still to be unlimited. Even—for example—if the values of the strong, weak, and electrical forces were very narrowly critical for the formation and stability of biological molecules and tissue, a similar necessity, for the gravitational constant, would seem to be unlikely.

More importantly, however, eliminating possible worlds by showing that they fail "anthropically" is considerably beyond our present knowledge of bio-molecular physics. The Anthropic Principle, therefore, even as a legitimate boundary condition on any string theory framework, seems, at least for now to be of unlikely usefulness for achieving the uniqueness of string-theoretic content necessary to describe our particular universe.

9

Beyond Quantumization

A Deeper Incorporation

Quantum supplementation (quantumization) in contemporary physics has made possible the framing of field theories that are not only capable of spectacular precision in predicting phenomena, but also have the ontological content required for explanation. Quantumization, however, is not without methodological awkwardness. Quantumized theories (e.g., quantum field theories and string theories) simply adopt, that is, conjoin, QT as a general "context" making use of its rules as needed for theory elaboration. This approach, however, subject only to the dictates of consistency, leaves much to be desired. Essentially, it is no more than a conjoining of some explanatory theoretical structure, H, with the rules of QT. It is therefore severely short on internal unity and leaves all the quantum perplexities entirely unresolved. It might seem, therefore, that a deeper incorporation of QT into any dynamical theory, H, would be one in which the formalism of QT is embedded *in* (rather than simply conjoined *to*) H, meaning by this that H (without QT) would logically entail QT. The content of H, of course cannot be exclusively that of classical physics, since we cannot recover QT from classical physics, which is incompatible with quantum physics. The content of H, therefore would have to include appropriate hidden variables; that is, H would have to be some sort of hidden variable theory.

As we have already seen, however, the difficulties incurred by such theories are serious, including even the impossibility of extending a strongly non-Boolean (noncommutative) structure of QT to the Boolean (commutative) one of classical theory. The derivability of QT from H therefore is blocked in principle. H, as a general theory of nature (micro, meso, and macro), would therefore have to go it alone, that is, without entailing aspects of QT. As a radically new theory, it would be a total replacement of QT—this time, however, an explanatory one.

Understanding and testing what H says about nature, however, requires extracting (deducing) some of its implication. This means devising some well-formed mathematical equations that express it and then finding some testable solutions of the equations. Now, though H would have to forfeit QT, it would still have to recover quantum statistics (facts of nature) and this could, in general, require differential equations that depart from linearity and even homogeneity—a huge and possibly dire loss of formal simplicity.[1] Bohm was able to do this for very simple systems. Doing so in general, however, and, depending on the extent of departure from linearity or other simplicity features, could call for techniques far beyond all present-day mathematical know-how. Science history, however, shows that the need for mathematical methods not yet available does not have to pose an ultimate impossibility. For, despite the daunting proportions of any challenge, the likelihood of mathematical discovery along the required lines—either in exact analysis or in approximative methods—cannot be dismissed entirely for the foreseeable future. Indeed, physical science and technology are brilliantly studded with monumental mathematical breakthroughs by physicists and natural philosophers or with novel applications of previously unnoticed, arcane portions of pure mathematics—all to meet the needs of physical theory.

So, though a philosophically more satisfying replacement of QT remains highly unlikely, its impossibility has never been conclusively shown. An explanatory theory that treats of an ontologically stable microworld, is consistent with the definiteness of the measurement experience, and possibly even accounts for the macroworld we actually observe—and, we might also add, does it all within the constraints of "metaphysical wisdom"—need therefore not be prospectively ruled out.

Nor is the replacement of QT utterly unthinkable. Despite its deep entrenchment on hugely confirming evidence, QT always remains subject to the possible interdictions of experimental findings. Like all other physical theory, it cannot be made necessary or eternally finalized by current observational confirmation. What is eternal about QT, or any other theory, for that matter, is not its "truth" but the real possibility of its emendation or even its entire replacement in the face of resistant puzzles, experiential facts left unexplained, or new unfavorable evidence. In this regard we might, for example, note recent astronomical findings that pose some challenge for QT only to remind us of its empirical vulnerability.[2]

More generally, however, there are significant limitations in the scope of QT. It does not, for example, predict the experimental values of any of our physical constants, for example, the (electrical) permissivity and (magnetic) permeability of free space, Planck's constant, the fine structure constant, the gravitational constant, the cosmological constant, and so on. Nor does it predict the existence of the basic forces of nature or of the bewildering swarm of diverse entities that make up the standard model of particle physics. And, as might be expected, it says nothing about the existence and interrelationship of inertial and gravitational mass, or about the equivalence of gravitation and acceleration, or about mass-energy equivalence, or, if you like, the emergence of consciousness in physical nature.

And there is more to motivate the necessary research. Recall the century-long foundational issues about the intelligibility of the quantum world and about how to link it dynamically and epistemically with the macroscopic world in which we make measurements to test the predictions of QT. Recall also the senses in which QT may be said to be "incomplete": It describes statistical aggregates rather than individuals; it fails on the definiteness, existential continuity, and (according to some) the very objectivity of substantial reality; and most notably, it lacks a sufficient interpretation for explaining what is going on behind the experimental appearances. As we noted early in our discussion, QT was born without an explanatory ontology a century ago and has yet to find itself one—the stubborn resistivity of its formalism being, arguably, fundamental. In the general context of incompleteness, we might also note the basic incompatibility of QT with relativity. QT is probabilistic and, on sub-microlevels (10^{-33}cm and 10^{-43} seconds), it models reality turbulently and discontinuously, while relativity is determininistic, and, at all levels, models space-time smoothly. And though this discrepancy draws little urgent attention, it remains a serious embarrassment for modern physical theory.

Finally, to all this one may add the other motivating foundational issues we have already discussed, the deepest and most pressing of which are the measurement problem and the closely related one of how to recover macrophenomena from QT. These remain essentially unresolved issues, with some of the most eminent efforts resulting in expansions of cosmic reality and in sweeping subjectivizations of experience that stretch to the unbelievable and beyond.[3] Motivation for reformulation and change therefore is not wanting as evidenced, for example, by the continued interest in nonlinear formulations of quantum theory, despite all the mathematical impasses.[4] Still, the alternative nearest at hand and most promising of foreseeable results is quantumization. For both scrutinizing and dancing with nature, it is presently an ongoing deeply established and rewarding strategy.

Conclusion

The striking mathematical elegance of QT and its amazing predictive success over a full century have made it theoretically central in virtually all modern physical contexts. Despite its greatness as a truly monumental achievement of our scientific culture, however, this theory is incomplete from the viewpoint of what a scientific theory, at best, is expected to do. It is incomplete in still lacking an ontology coherent and structured enough for explanatory purposes. For, though QT predicts with stunning precision over a vast range of phenomena, it provides no explanations of what it predicts, either in a causal (strong) sense or in a unifying (weaker) sense of explanation.

This need not be very surprising, for almost entirely throughout its growth, QT was, for the most part, crafted not for explaining but for computationally smoothing over some embarassing discrepancies in the going physics and for predicting a relatively narrow though compelling set of phenomena left behind by classical theory. The result was several formalisms with structures so starkly algorithmic (matrix mechanics) or so abstract (matter-wave mechanics) as to resist not only a generally acceptable interpretation but also its very possibility.

One, however, can still ask the question that has rankled the foundations of quantum theory from the very beginning and throughout: How can an abstract algorithmic scheme such as QT yield predictions that are so astoundingly accurate and yet so suggestive of an underlying reality that is abstrusely unintelligible or, at best, stubbornly counterintuitive? This is the question that has fueled the endless (if not obsessive) search for the appropriate interpretation of QT or, as one might say, for an adoptive ontology that can make some sense of it all. The predictions of QT—at least as they are generally construed—raise gnawing questions about the existential stability, determinateness, and even objectivity of microreality. What is more, the linkages (entanglements) that QT features for such reality do not fit the general standards of causal intelligibility. Finally, there seems to be no broadly acceptable manner of resolving the measurement problem, that is, of reconciling the indefinite cluster of informational possibilities that the core quantum formalism delivers with the phenomenological definiteness of actual observational experience and, derivatively, of the entire macroscopic world.

These questions are all issues that bear directly on what scientists believe science is all about. In their philosophically unguarded moments, scientists tend to be objective realists, and they seem to feel this credo very deeply, as evidenced in most of what they do and say when they theorize, experiment, or report their results. In effect, they believe (and, it seems, with all their hearts) that science must ultimately be about a reality that is observer-independent, space-time determinate, and linked by local causes. Otherwise, it would be little more than the inventing and applying of alchemic, magic-like, symbolic devices for prediction only.

In its present form and despite its awesome elegance and breathtaking scope, QT does not escape the pressures of this requirement. In the final analysis, the abstract formalisms (matrices and wave equation) in terms of which QT is standardly expressed are algorithmic constructions offering virtually no structural clues about how to construe the underlying dynamical subject matter and providing, therefore, little guidance toward any solidly explanatory interpretation.

Still, the interpretational program has engaged many distinguished theorists and has resulted in a wide variety of "interpretations," all remarkably and ingeniously crafted and elaborated, yet all failing general acceptance. Among the various foundational problems, some of the most puzzling or controversial have been wave-matter duality, irreducible uncertainty and indeterminacy, incompleteness in various senses, subject matter objectivity, quantum non-locality, the measurement problem, the unexplained emergence of classicality from the quantum substrate, and finally, even the role of consciousness in the evolving quantum world. But despite the remarkable creativity in the addressing of these issues, the resulting interpretations have all remained deeply controversial, leaving issues inconclusively treated, or, when making some headway, doing so at a heavy cost of theoretic virtue. Indeed some of the most recent and elaborately formulated interpretations have typically resorted to ontologically inflationary creations that critical responses have either scored on physical significance or charged with out and out metaphysical turpitude. Indeed, some of these have been further characterized as wildly imaginative, unbelievable, or even extravagant.

The historically dominant and ever-competing interpretations of QT are several, and they span a wide range of ontological approaches, all problematic. CI, for example attempts to avoid the paradoxes of wave-particle duality and of the disallowed simultaneous measurement of noncommuting variables by flatly denying the existence of all observables except during measurement. Remarkably, versions of this foundational philosophy have found wide acceptance among leading physicists (at least, in their less realistic and more speculative moments). It nevertheless remains subject to a range of serious objections. Recall, especially, that CI thins out QT's ontology to the point of depriving it of explanatory power, consequently allowing quantum weirdness to linger as a brute fact of nature. Further, it needs the von Neumann-Born measurement postulates (not derivable from the core formalism) in order to allow satisfactory prediction in quantum measurements. And further still, it fails entirely to resolve the measurement problem, to boot, leaving the quantum and classical worlds completely sundered.

HVI, on the other hand, postulates "unobservables" that the prevailing positivistic or operationally inclined theorists consider much too low on physical significance and therefore metaphysically burdensome. What is more, hidden variable theories must be able to recover the statistical predictions of QT. The findings of various investigators, however, pose logical impasses (so-called "no-go theorems") to defining physical states in terms of hidden variables with uniformly determinate values over which one can average in order to recover the quantum statistics. Further, in attempting such reduction, nonlinear terms may have to be introduced at a steep price in mathematical simplicity and workability. More at odds with the realism of HVI, however, the postulated hidden parameters turn out to be non-local, therefore in violation of common sense, and arguably contrary to the spirit, if not the letter, of special relativity.

Von Neumann's projection postulate, along with other collapse theories—though better regarded as supplementations of the core quantum formalism rather than as full-fledged hidden variable interpretations—raise serious issues of their own. Wave collapses are discontinuous non-local processes, and they give rise to the gnawing issue of where to place the "cut" between deterministically evolving (multiple) superpositions and the abruptly definite ("unitary") outcome experienced in measurement. So stubbornly pressing was this problem that it led von Neumann and, later, E. P. Wigner[5] to subjectivize all that we observe experimentally, that is, regard it as mind-dependent. They concluded that collapse can occur only in the presence of consciousness, to wit, only when someone experiences an outcome of measurement. Moreover, this view must be dualistic. That is, consciousness must be taken to be something nonphysical, for, were it to be construed materialistically, as part of the material world, the problem of where to place the cut would have to persist.

At any rate, the introduction of consciousness in this manner leads to a subjectivization of the physical reality we believe in on the basis of our perceptions. It also leads to the belief-defying conclusion that where there is no consciousness there are only superpositions. So that, until a conscious entity observes, some organisms are both dead and alive; one and the same object can simultaneously "exist" in many places; and the pointer in someone's experiment can simultaneously "point" to more than one setting. Indeed, on this view, before there were conscious beings, the world must have consisted not of anything definite (discrete) but of a vastly convoluted superposition—a "reality" whose structure seems to transcend anything imaginable or otherwise intuitable.

Other mainstream interpretations of QT do no better. Testing the limits of reasonable belief even further than its predecessors, for example, are the many worlds type of interpretations (MWI), which attempt to recover both the unitary discreteness (definiteness) of the experienced outcome of any measurement and the experimentally correct distributions (quantum statistics) observed in a large number of quantum measurements. To explain observational discreteness, MWI postulates a superposed multiworld of countlessly many coexisting discrete worlds that evolve and branch according to the Schrödinger equation and that therefore mirror the full mathematical elegance of QT's core formalism. Then, to recover quantum statistics, the ontology is further inflated by assuming that the number of postulated worlds is infinite and capable of accommodating the needed probability measures.

An unyielding stumbling block for this bloated ontology is the lingering need of a preferred basis. Such a basis cannot be derived from the core formalism and must therefore be assumed as an outright add-on in order to select from the vastly superposed set of alternative states. Adding bottomless metaphysical depths to its inelegance, MWI shields the entire multiworld from all detection—except for the tiny portion that the observer happens to inhabit at the moment. It seems hard to imagine even the most metaphysically tolerant theorist as not wincing here. The almost entirely hidden world casts a deep metaphysical shadow on the empirical status of the overall ontological framework.

The pressure on MWI does not end here. Any sense of personal biography becomes unimaginably difficult to untangle, if not utterly compromised. How do we register an individual's personal identity if she/he shares her past with many others at forks in the branchings, or if, in the very next moment, her/his different quantum states will respectively inhabit other worlds? The ad hoc "tunnel vision" hypotheses with which MWI theorists attempt to cope with this mind-numbing perplexity is another add-on that weighs heavily on theoretical virtue. At this point, it would not seem unreasonable to charge MWI with enough extravagance to stretch credulity to the breaking point.

The many minds interpretation (MMI) is a response to these issues, but meets with serious difficulties of its own. It rejects MWI's gratuitous assumption of a natural, objectively real preferred basis and associates with each observer in the smoothly evolving multiverse a simultaneous multiplicity of "minds" each having its own experiential content. A measurement therefore results in a whole set of alternative states of awareness for each observer, and this, of course, still leaves us with the question of why actual experience is always discrete (singular, coherent).

MMI's amazing answer is that the definiteness of "actual" experience is only an illusion of consciousness whose explanation is not the burden of QT but, possibly, of some future theory of consciousness. Such a sidestep, however, severely compromises the validating role of observational confirmation (or falsification) in physical inquiry and, for this reason alone, cannot be taken seriously. What is more, bringing states of consciousness—illusory or not—into a physical theory would seem to require that they be physical and therefore superposable. This, however, would clash with the facts of introspection which reveal memories (results of measurements, etc.) that are quite accessible, quite definite, and basic to all personal identity. To avoid this predicament, some MMI theorists opt for dualism by taking minds to be "something other than physical."

This version of MMI, however, raises other serious issues. Like MWI, in order to recover quantum statistics, it postulates a continuous infinity of worlds. Unlike MWI, however, it associates with each brain state in the evolving superpositions of measurement outcomes a continuous infinite set of non-physical and therefore non-superposed entities or minds. Sets of non-physical minds therefore respectively index or "track" successions of alternative brain states in the evolving composite physical system. Along the way, "individual minds" will randomly "choose" membership in one or another of the indexing infinite sets, so that the resulting process is genuinely stochastic and, under appropriate assumptions, demonstrably capable of yielding the correct quantum statistics.

While this scheme avoids some of the embarrassments of MWI, it incurs others. For whatever the objection is worth—infinite sets of minds can be shown to be weakly non-local. More seriously, however, this dualistic version of MMI uncritically dispenses with the important mind-body relation of supervenience which requires a tight relationship between brain states and mental (i.e., subjective) ones and which is so central to any current dualistic philosophy of mind. The scheme consequently leaves us with no physical earmarks whatever for telling which individual minds end up tracking which alternative outcomes of measurement in the evolving system. The metaphysical status of non-physical, "supervenience-free" minds, therefore, remains deeply mysterious and subject to all the traditional difficulties that roil the mind-body issue.

To sidestep the supervenience requirement and the related mind-body issues some versions of MMI opt for reverting to a strictly materialistic ontology in which the "composite mind," consisting of the many minds associated with a brain, is itself physical as a subsystem of that physical brain. But, though such a full slide to materialism might seem desirable from the viewpoint of physical theory, it can be seen as precipitous. By leaving subjective states completely out of the picture, it presents an incomplete and philosophically unsatisfying ontology for anyone who believes in the truth of dualism and in the reality of qualitative subjective states. Some materialist proponents of MMI have therefore, responded by allowing, in some metaphysically foggy sense, the "subsistence" of the "quality" of an otherwise "physical experience." Unfortunately, however, these accounts leave much to be desired on whether, in the final analysis, this quality is itself physical or non-physical.

A materialistic MMI raises still other serious questions. For example, if consciousness consists of simultaneously multiple physical states, are the corresponding multiple experiences also simultaneous? Indeed, some MMI theorists (it seems, grudgingly) concede that experiences must occur at differing discrete times rather than simultaneously, or there could be no personal identity—such identity being rooted in a discrete set of past memories. But then, how does consciousness "tunnel" through the manifold of simultaneous experiences to yield a definite and discrete personal biography? There is an acute need, here for an underlying theory of consciousness that can deliver the required details. Absent such a theory, this and other related questions about the workings of consciousness in MMI remain essentially unanswered.

No less troubling, for both MWI and MMI is the more general matter of theoretic virtue. In their attempt to recover quantum statistics, avoid wave collapse, settle on the preferred basis issue, and deal with the facts of consciousness, these attempts at making sense of QT pay dearly in elegance. They inevitably find themselves elaborating a vast complex of inaccessible worlds comprised of continuous infinite sets, somehow laced with shadowy minds that play according to psychophysical and "intra-psychic" laws having no theoretical basis. One might well ask here: Does any of this make for better theorizing than flatly adopting the Born and von Neumann postulates?

More recent history has seen the rise of decoherence theory (DT) as a fresh approach to some of the issues left unresolved by the available interpretations. DT represents an interesting and, one might say, "on the ground" attempt to address foundational issues such as the measurement problem and the emergence of classicality from the quantum states that presumably underlie all physical existence. It does this by conferring some dynamical properties on quantum systems and by universalizing the quantum nature of all entities—"micro," "meso," and "macro." Moreover, in response to the problems of measurement and recoverance, it invokes the notion that the measurement interaction and the resulting entanglement induce, within the superposition that the formalism delivers, a "natural selection" of "preferred" or unentangled, interference-resistant subset of states. These are then viable candidates for the definite macroscopic measurement result. There is, however, nothing in the formalism to suggest any of this, so that the idea is ultimately a theoretically extraneous add-on. Furthermore, the account leaves us with a multiplicity of decohered alternative outcomes indicating a need for further interpretation—possibly of the many worlds type. This, however, would take DT back to the quandaries of MWI.

Also within more recent history GRW theory approaches the measurement problem with a vigorous attempt to explain wave collapse in terms of certain highly specified dynamical interactions. Despite its intriguing explanatory aura, however, GRW raises a serious question of intelligibility. Its account pivots on the mysterious notion of a physical state "interacting" with an abstract wave function whose physical significance defies all definition. This is a metaphysical gap that blocks explanatory function and leaves the algorithmic status of QT essentially unchanged.

Finally, there are what we have characterized as formal interpretations and what some have called "quantum logic." These attempt to enlighten the formalism of QT by expressing it in terms of an algebra, logic, or other computational system. But while these schemes have done much to clarify the quantum-theoretic formalism and enhance its range and applicability by taking it to some higher level of generality, they have had to fall short on at least some of the goals of interpretation. Such formal transformations are not ontic interpretations and therefore do not provide the existential subject matter or ontological base a physical theory needs in order to "qualify" as an explanatory nomological.

Well—the very long search for an acceptable interpretation of QT has been a creatively intense polemic that, despite its historic proportions, has resulted in no generally settled body of opinion. The ongoing highly resistive difficulties that have plagued the program from the very start have led to some of the most imaginative creations in the history of scientific ideas. These have generally had to be either highly elaborate but problematic constructions or drastically diminished ones with unstable or vanishing ontologies. Or they have had to be last-measure retreats to subjectivisms and observer dependence that do not sit easily in the traditional naturalism of physical science.

These are results that caution demureness on the problem of interpretation. On simple pragmatic grounds, at least, there seems to be reason to relent on the entire program. Indeed, the formal obstacles to any ontic interpretation adequate for explanatory purposes threaten the goals of the program even on the very level of sheer possibility. The fresh formulation of new theory, adequate in formal structure and rich enough in ontological content (metaphysical qualms aside) to explain, rather than merely predict, natural phenomena, therefore, would seem to be the more promising investment of theory-creating effort. What is more, the formulation of basically new theory is not only a live possibility but one that has already delivered some highly successful explanatory accounts of the microworld, to wit, quantum electrodynamics and quantum field theory.

Such theories neither jettison nor recover (as HVI would have it) any aspects of QT. Rather, they are formulated as hybrids, that is, as explanatory theories elaborated in the context of the quantum-theoretic rules. Quantum field theories assume the existence of point-particles interacting under four basic forces brokered by messenger particles. Further, this dynamical ontology includes antiparticles for each of the fundamental particles and also (thanks to QT) allows for the existence of spin. These theories provide highly successful explanatory bases that yield the standard model for particle physics, and much more. Similarly, string theory with its ontology of string loops, branes, and multidimensional spaces promises explanatory theory of unprecedented unifying power.

These theories, however, do all this in the context of QT. The result is quantum-type or quantumized theories that provide appropriate quantization along with other quantum rules and allowances such as uncertainty, exclusion, negative energy and the paroxysmal creation and annihilation of dynamical but transient (virtual) particles on returnable cosmic energy. This is theorizing, as it were, with "quantum tolerance." The adoptive quantum-theoretic features, though not themselves explained within the new theory, can provide considerable predictive leverage. We might also add here that, though QT is algorithmic and, by itself, ontologically insufficient for explanatory purposes, it can, as a context for other theories, help augment their ontology and therefore their dynamical power. For example, it can, with probability-governed "allowances," permit the ontological expansions—albeit some only transient—that can be vastly effective for explaining the basic force interactions.

In quantumized theories, quantum features such as uncertainty, non-locality, and the overall granularity of nature, could remain bottom-line aspects of the natural world—at least for the time being and always within the dictates of consistency and bottom-limit intelligibility. Quantum non-locality, for example, is not, strictly speaking, a genuine lapse of causality. It is only an increased statistical correlation between separated parts of an entangled quantum system. If there is randomness in one part of the system, there will still be randomness in the other part of the system, so that no informative signal can be sent instantaneously. Quantum non-locality therefore remains relativity friendly. The apparent "influencing" at a distance may be less congenial to the understanding than causation by contiguous links. But such *influence* relations are at least analogous to *causal* ones, though unmediated or "non-local," and therefore are easily intuited. After all, we experience many causes whose effects appear to be instantaneous actions at a distance. So, objective realists would not have to summarily reject quantum tolerances. Advisedly, they could most probably live with a liberalized philosophic stance in which commonsense (objective) realism gives way to a more permissive, but highly effective critical realism.

Meanwhile, there still beckons the possibility of an overall replacement of QT with a theory of nature that would make for a fuller mathematical integration of contemporary physics by recovering—*through explanation*—the predictions of present-day quantum and quantumized theory together with all that classical theory (relativity included) can tell us about the macroworld. This is a possibility that, though perhaps distant, is no less real than was the possibility of an Einsteinian reformulation of Newtonian dynamics. Indeed, it is a possibility real enough to offer a long-term focus for current and future research in theoretical physics.

Notes

Chapter 1. The Quantum and Classical Theories

1. A black body is one whose surface is a perfect absorber of electromagnetic radiation. Such a body will also be an optimal emitter of radiation.

2. What we are calling quantum-type or "quantumized" theories are physical theories that have been developed in the context of (while joined conjunctively to) quantum theory, or, as we might say, that assume the correctness of QT. The term *quantumized* is not synonymous with the term *quantized,* which designates a particular aspect of quantum theory. Examples of quantumized (i.e., quantum-type) theories are the quantum field theories and string theory.

3. Locality, a constraint that we will visit later, roughly speaking, requires that no system can *instantaneously* influence any system spatially separated from it.

4. Somewhat infrequently noted in this general regard is the fact that the familiar observational content (scintillations, clicks, trajectories, etc.) picked out by the vocabulary of classical physics remains operative in QT via Bohr's *correspondence principle,* which urges not only that quantum mechanics be, in certain general respects, consistent with classical theory, but that, in limiting cases, it yield the classical laws to a suitable approximation. This principle has been a powerful guide both heuristically and technically in the overall development of the subject as, for example, in discovering some of the quantum selection rules.

5. The Hamiltonian, H, of a system is the total energy of the system expressed as a function of the spatial coordinates and momenta of the particles in the system. Therefore (in rectangular coordinates), for a system consisting of a single particle of mass, m, moving with momentum, p, through a field of potential energy, $V(x,y,z)$, $H = 1/2m\ (p_x^2+p_y^2+p_z^2)+V(x,y,z)$.

Chapter 2. Quantum Puzzles

1. De Broglie conjectured that if electromagnetic waves were also particles (photons) then matter particles were also waves (matter waves).

2. More formally, Ψ is understood to be a vector, $|\Psi\rangle$, in Hilbert space that, when used in accordance with the rules of QT, can generate sets of numbers that are interpreted as referring to the probabilities of possible outcomes of measurement. With each observable, L, is associated a Hermitian operator, *L*, with a complete set of eigenvectors, $|L_1\rangle,|L_2\rangle$ . . . , such that the possible results of measurement of L are restricted to a set of discrete values. i.e., the eigenvalues, L_1, L_2 . . . , of L.

When a quantum system is moving without confinement, basically, two rules determine the way $|\Psi\rangle$ evolves for the system, given any initial $|\Psi\rangle$: (1) When the system under consideration is not being measured, the rule by which it evolves is the time-dependent Schrödinger equation: $(-h^2/8\pi^2m)\ \nabla^2\Psi + V\Psi = -h/2\pi i\ \partial\Psi/\partial t$. (2) When and only when the quantum system is measured, the rule that applies is that the probability of a transition from $|\Psi\rangle$ to $|L_i\rangle$ is the square of the projection of the vector, $|\Psi\rangle$ unto the vector, $|L_i\rangle$, i.e., $|\langle L_i | \Psi \rangle|^2$. The result of a measurement of the observable, L, is understood to be L_i, where L_i is the eigenvalue corresponding to the eigenvector, $|L_i\rangle$ of L, and $|\Psi\rangle$ is projected (collapsed) unto the vector, $|L_i\rangle$ (The expectation value of L is given by $\langle L\rangle = \int\Psi^*L\Psi d\tau$ where $d\tau$ is an element of space and the integration is over all space.).

When a quantum system is confined to a certain region, $|\Psi\rangle$ obeys the time-independent Schrödinger equation, $(-h^2/8\pi^2m)\ \nabla^2\Psi + V\Psi = E\Psi$ where E is the total energy of the system. The particular solutions of this equation have the property that $\Psi\Psi^*$ and therefore all probabilities are constant in time, varying only over space. Accordingly the solutions are said to represent *stationary states.*

3. The spin of some quantum particles corresponds to the possibility of polarizing the associated "waves."

4. Generally speaking, in vibration theory it is convenient to assume complex solutions that vary exponentially with time. This generally simplifies the calculation. To obtain physically significant results we then use the real part of the function we come up with, which is also a solution. The Schrödinger equation, however, which contains not only second (space) derivatives but also a first (time) derivative, admits only of solutions that vary as complex negative exponentials whose real parts would *not* be solutions. This immediately poses a problem in any attempt at interpretation.Ψ is inherently and irreducibly complex; yet, as it seems, it must be regarded as representing some physical quantity.

5. For a system of N charged and therefore interacting particles, Ψ is a function not merely of the coordinates of one particle but of the coordinates of all N particles. (Intuitively, this complication is suggestive of interactions between all particles in the system.) Ψ therefore "moves" in a 3N-dimensional space.

6. The references in the literature to the time-dependent Schrödinger equation as an "equation of motion" are too numerous to require citing.

7. More precisely, the probability of finding a quantum particle in the volume element dxdydz is $\Psi(x,y,z)\Psi^*(x,y,z)\ dxdydz$.

8. A phase space is one used for later formulations of Newtonian mechanics in which the coordinates of a dynamical system are specified as position and momentum.

A non-Boolean logic or "algebra" is one for which some of the rules of standard (Boolean) logic do not hold, as for example, the commutative rule for conjunction by which, for proposition A and proposition B the proposition A and B is logically equivalent to the proposition, B and A.

9. For an elegantly detailed account along these lines, see Jeffrey Bub, *Interpreting the Quantum World* (Cambridge: Cambridge University Press, 1997).

10. For a similar but well-illustrated popularized account of the cosmic delayed action experiment, see physicist N. Herbert's *Quantum Reality,* (New York: Anchor Books, 1985), 164–67.

11. Very generally expressed, Bohr's complementarity principle says that quantum subject matter has some essentially dual aspects and that therefore any complete description of it must be given in two complementary ways. Understood in this general manner the principle has also been expressed with respect to the "complementary" nature of the paired (conjugate) observables p and q (momentum and position) and E and t (energy and time). These quantities expressed as state variables are fundamental in the description of any quantum system. The operators corresponding to the variables in any such pair, however, are noncommuting, which makes the variables fundamentally subject to Heisenberg's uncertainty principle thus deeply linking them, while in principle limiting the precision with which they can be simultaneously determined. Non-commutativity is a differentiating or essential property of quantum operators corresponding to conjugate variables. This is a specifically quantum property, i.e., one that is not attributable to classical variables.

Bohr related this indeterminacy to the necessary incompatibility of the experimental setups needed for the simultaneous measurement of "conjugate" observables. But the experimental conditions and the quantum subject matter they make it possible to observe were for Bohr a *total* physical system in which the indeterminacy linkage of conjugate variables reflected the "complementarity" necessary for a "complete" description of what is being observed.

12. The time dependent Schrödinger equation is linear because it does not contain Ψ or its derivatives in powers greater than one and homogenous because it does not contain any terms independent of Ψ. See Note 2.

13. More technically, this means that the wave function, Ψ, can be written as the expansion (superposition) $\Psi = \sum_i c_i \psi_i$ where the ψ_i are linearly independent orthogonal eigenstates of some dynamical variable and the c_i are complex coefficients.

14. Superposition is not to be thought of as a consequence of only the Schrödinger formalism. It also arises in the equivalent Heisenberg matrix formalism. Indeed, the matrix formalism is especially suited for elaborating the formal features of quantum mechanics when a dynamical variable can take on only a finite number of values, as, for example, in the case of spin,

a "dichotomic" or two-valued variable. In such a case, if the possible spins are +1 and –1, we can represent the spin state, χ, of the system by the one-column matrix or *spinor* whose elements are $\chi(+1)$ and $\chi(-1)$, and which can be expanded into the *superposition* $\chi = c_1\alpha + c_2\beta$ where α and β are particular (possible) spin states, $c_1 = \chi(+1)$ and $c_2 = \chi(-1)$. The spin state, χ, can be regarded as a vector with complex components, c_1 and c_2, in a two-dimensional linear vector space. The expressions, $|c_1|^2$ and $|c_2|^2$ are the probabilities of finding the particle with spin up or with spin down, respectively.

15. The puzzle of superposition intensifies still more with the fact that some of the possible states represented by the evolving wave function can turn out to behave strangely enough to exceed the initial energy of the system. One of the constituent particles, for example, may suddenly relocate and find itself entirely outside the energy barriers of the system, as in the phenomenon of *tunneling*. What is disquieting here of course is the threatened violation of one of the most deeply entrenched symmetry principles of physics, namely, mass-energy conservation.

16. I owe this simple example to a very similar one in, H. Putnam, "A Philosopher Looks at Quantum Mechanics," in *Mathematics, Matter, and Method: Philosophical Papers*, Vol.1 (Cambridge: Cambridge University Press, 1979), 2nd Ed., 134.

17. A "basis" is a linearly independent set of vectors, $\psi_1, \psi_2, \ldots \psi_n$, that may be chosen from among the vectors in an n-dimensional Hilbert space. Any vector, Ψ, belonging to such a space can be expanded (decomposed) in terms of these *basis vectors* yielding the superposition $\Psi = \sum_1^n a_i \psi_i$ where the coefficients a_i are complex components of Ψ in that they determine Ψ completely. In the present context, Ψ represents a quantum state "evolving" in the n-dimensional Hilbert space.

For purposes of useful theoretical elaboration it is important that a basis be both orthogonal and normal. This means that the vectors of the basis must be mutually orthogonal (i.e., the corresponding inner products must be zero) and they must be all of unit length. Orthogonality assures the linear independence of the vectors. The normality requirement is needed to satisfy the probability concept. Orthonormal sets of n such basis vectors can always be constructed.

18. Other formal difficulties arise with the assigning of probabilities to alternative outcomes. Though the amplitudes of the superposed possibility waves add up linearly, the probabilities of superposed possibilities do not—as would classical ones. The problem, of course, arises because the probability associated with each component (of the superposition) is proportional to the square of its amplitude. Without proper remediation, therefore, superposition predicts what appear to be incompatible probability percentages for the occurrence of mutually exclusive observables in a system consisting of a linear superposition of possible quantum states.

19. H. D. Zeh notes that superpositions of such entities as photon pairs, fullerene molecules, and macroscopic currents running in opposite directions have been claimed to exist as physical states by various investigators.

But he also concedes that such existence is still not acknowledged by many physicists. In any case, it seems impossible to imagine what the simultaneous and observationally direct "experience" of more than one component would be like; such a thing would seem to be phenomenologically impossible. The simultaneous "occurrence" of such states would, it seems, have to be some sort of experimentally indirect (inferential) determination.

In the same discussion, Zeh does point out that when the appropriate interference experiments have been properly done all the components of a superposition have been *indirectly* shown to exist simultaneously. This, however, entails rejecting those interpretations of QT that allow existential status only to what is *actually* observed. See: "Basic Concepts and their Interpretation," in E. Joos et al., *Decoherence and the Appearance of a Classical World in Quantum Theory* (Heidelberg: Springer-Verlag, 2003).

20. Investigators have attempted to give certain kinds of superpositions physical meaning by assigning interpretations based either on other principles (quantization rules) or on experiments. Along these lines, it has been maintained that such superpositions "exist" in nature. Even in this "interpretational" sense of "existence," however, it is generally believed that not all superpositions can be found in nature. Indeed, superpositions of protons and neutrons ("preutrons") have never been observed. Accordingly, theorists have formulated "superselection rules," either by postulating them outright or by deriving them from quantum-theoretical contexts such as quantum field theory. Though some theorists have even sought to implement decoherence theory for the justification of such rules, a theoretical basis for them would require a modification of the Schrödinger formalism to allow there to be some observables that commute with all observables. Absent such a modification, superselection rules remain as essentially ad hoc, theory-independent supplementations. See A.S. Wightman, "Superselection Rules; Old and New," *Il Nuovo Cimento* 110B (1995): 751–69.

21. In standard QT, the way out of this is postulational and statistical. The Schrödinger formalism does not, by itself, provide the desired unitary outcome of the measurement interaction. But with the help of additional assumptions such as the Born probability interpretation and the von Neumann projection postulate, one can speak of a "wave collapse" in which an ensemble of possible measurement outcomes gives way to a single alternative, observable as a *definite* experience. On the basis of the Born postulate, we can compute the respective probabilities that this or any other outcome will actually materialize.

22. More precisely, if the eigenvectors corresponding to the eigenvalues of an observable to be measured in a system, S, are denoted by s_k and those corresponding to the pointer position of the measuring device, A, by a_k, then the state of S before its interaction with A is given by the superposition, $\Psi = \sum_k c_k s_k$ and the state of the composite system after S interacts and couples with A is given by the composite superposition, $\Psi' = \sum_k c_k s_k a_k$. When the observation is made, however, what is observed is a definite pointer reading, p, indicating that the state of the composite system is $s_p a_p$. The radically urgent issue here is that of explaining how the superposition of alternative and experientially

incompatible states (i.e., $\sum_k c_k s_k a_k$) becomes, upon observation, a definite state, given by $s_p a_p$.

23. The projection postulate is *not* derivable from the core formalism and has no other nomological basis. The process it posits is discontinuous, irreversible, indeterministic, and—in a sense to be explored later—"non-local." As a law governing the interaction of a quantum system and a measurement setup it is therefore of a very different sort from the quantum law (i.e., the Schrödinger equation) governing the evolution of an unmeasured quantum system. Purportedly, however, both these processes are entirely physical. Therefore, that they should require two separate "physical" laws—moreover, of radically different forms—is methodologically awkward, if not flatly unacceptable. Certainly we have here a vast diminution of elegance.

24. The idea of placing this "von Neumann cut" in the observer's consciousness had already been entertained by Heisenberg. Later it was also supported by Wigner and others. Wigner explicitly allows that a conscious state can cause a physical event. See: E. P. Wigner, "Remarks On the Mind-Body Question," *The Scientist Speculates, ed.* I. J. Good (New York: Basic Books, 1961), 284–301.

Chapter 3. More Quantum Puzzles

1. More generally still, such theorists have attempted to make sense of (i.e., interpret) QT in terms of a purportedly "deeper" or "hidden" subject matter that conforms to commonsense realistic requirements such as objectivity, determinateness, mediated (local) causality, and so on. These attempts are what we later collectively discuss as the *hidden variable interpretation.* In a hidden variable theory, the statistical predictions of QT would be recovered by averaging over the values of hidden variables.

2. The uncertainties in the conjugate variables, position and momentum, are formally related by the Heisenberg uncertainty principle, a theorem of quantum mechanics. For these quantities, the uncertainty principle may be written as $\Delta x \Delta p \geq h/2\pi$ where Δx is the uncertainty of the position and Δ p is the uncertainty of the momentum.

3. This is the solution for which $\Psi = Ce^{-2\pi i/h\,(Wt-px)}$ where p is the particle's momentum, x its position coordinate, $W = p^2/2m$, C is a constant, and h is Planck's constant. This means that Ψ cannot be normalized because $\iiint|\Psi|^2 dxdydz = \infty$.

4. In this case, the wave has only *phase* velocity and not group velocity.

5. More precisely, this means adjusting the constant, C, so that $\iiint|\Psi|^2\, dxdydz = 1$. This normalization requirement follows from the obvious fact that the particle is to be found with certainty, i.e., with a probability of one, within the wave packet, however large. This standard ordinary aspect of the probability concept is formally expressed in the probability calculus by the axiom that the sum of the probabilities of all possible alternatives is *one.*

6. In more formal terms, the wave packet, Ψ, can be expressed as a Fourier integral.

7. This expression, by itself, of course, clearly implies that neither the position nor the momentum can be determined with total accuracy. Zero error in one would result in a singularity in the other.

8. It is also demonstrable from the quantum formalism that since the components of the angular momentum, L, do not commute, definite values cannot be assigned to all its components simultaneously and an exact simultaneous determination of all three components of L can be made only if the expectation value of L is zero, i.e., $\langle L \rangle = 0$.

9. Vectors are quantities having both magnitude and direction. A change in direction, therefore, is change of quantity.

10. For a similar but more detailed and diagrammed account, see, J. C. Slater *Quantum Theory of Matter* (New York: McGraw-Hill, 1951), 25.

11. In such a case, a solution of the wave equation results in a wave packet, namely, a Ψ that has values appreciably different from zero only within some limited region.

12. When Heisenberg first formulated his principle in 1927, he did so in the context of wave mechanical considerations by expressing a complex "matter wave" as a Fourier integral. He certainly had no need of any disturbance model in any of his thinking.

13. E and t are noncommuting variables so that $\Delta E \Delta t \geq h/2\pi$, where ΔE and Δt are the uncertainty or error in the energy and time measurements, respectively.

14. I refer here to interpretations of the Copenhagen type.

15. The energies, masses, and charges in this quantum mechanical account give it some aura of explanatory content. For a full and genuinely causal explanation of electron-positron creation and annihilation, however, we must go to quantum electrodynamics (QED)—an explanatory and quantumized field theory of interactions of charges, radiation, and matter. We will pause for some brief attention to QED in our later discussion of quantumized theories.

16. A similar scenario can be envisaged for the creation of virtual photons, which can arise spontaneously out of the void but exist only on "borrowed" energy that they must return by going out of existence. The larger the energy, ΔE, of the virtual photon, the more egregious is its (temporary) violation of the conservation laws. Its debt to the cosmos is therefore more "urgent," so that its existential duration, Δt, is shorter. Similarly, a longer Δt means a smaller ΔE. (Mathematically, $\Delta E \Delta t = h$.) Reducing Δt may be understood as "locating" the photon more precisely in time.

17. Schrödinger, for example, took *entanglement* to be the feature that most distinguishes (separates) quantum mechanics from classical theory. E. Schrödinger, "Discussions of Probability Relations between Spacially Separated Systems," *Proceedings of the Cambridge Philosophical Society* 31 (1935): 555–63.

18. A. Einstein, B. Podolsky, N. Rosen, "Can a Quantum mechanical Description of Physical Reality be Considered Complete?" *Physical Review* 47

(1935): 777–80. This famous article is more easily available in S. Toulmin, *Physical Reality* (New York: Harper and Row 1970), 122–42.

19. For such a system of two particles, A and B, the two quantities, q_A-q_B (their separation), and p_A+p_B (their total momentum) are *commuting* variables. Quantum theory therefore allows their precise simultaneous assignment with no violation of Heisenberg uncertainty.

20. At first blush, this result invites the conclusion that the Heisenberg uncertainty principle has been violated and that therefore QT is inconsistent by both allowing and disallowing the exact determination of a particle's position and momentum. The defending response, however, is that Heisenberg uncertainty applies only to direct measurement and not to the indirect method of EPR. Einstein eventually did go along with this. But he insisted that QT is nevertheless incomplete, for it does not allow the existence of properties that EPR still shows to be "experimentally realizable" and therefore real.

It is worth noting in this regard that the defenders against the EPR charge of incompleteness were largely followers of the Copenhagen interpretation of QT led by Niels Bohr. According to this interpretation, observables blinked in and out of existence depending on whether or not they were being directly measured. The Copenhagenist rejoinder therefore was that unmeasured observables had no physical significance and that it was therefore meaningless to speak of the existence of any observable except during *direct* experimental measurement.

21. It is sometimes claimed as, for example, during a Conference on Quantum Mechanics and Reality at the Graduate Center of City University of New York (March 7, 1986) that simultaneous measurements of noncommuting variables can be made if you measure commuting variables that are coupled by a strong Hamiltonian to the noncommuters. Bohr's answer to this EPR-type claim would very likely be that such a measurement is not a "direct measurement." The rejoinder that followed, at the mentioned symposium, when I gave this kind of answer, was that no measurement is ever "direct" and that we bring concepts and theory into every measurement. Such a discussion may be seen as suggesting that measurement is itself a dangling concept in QT.

22. Mediated causality—a relation that cannot be instantaneous action at a distance—was basic to Einstein's objective realism.

23. Though on the standard model elementary particles are dimensionless points, QT allows them to have "spin," a property somewhat analogous to rotation but, of course, incoherent and therefore ontologically vacuous from any classical point of view.

24. QT does not provide for any measurement-independent existence of an individual particle's spin components. Indeed the orthodox and dominant Copenhagen interpretation (CI) commits to such an ontological gap by explicitly denying all intermeasurement ontology. This would in itself clash with Einstein's realism and inspire the charge of incompleteness.

For a recent resurfacing of the issue of compatibility between quantum entanglement and special relativity, see: D. Albert, "Special Relativity as an

Open Question," *Relativistic Quantum Measurement and Decoherence,* ed. H. P. Breuer and F. Petruccione (New York: Springer, 1999), 1. Albert's argument pivots around the claim that the quantum mechanical feature of *entanglement,* which violates the causal separability of events, clashes with the Lorentz invariance required by special relativity. In a forceful rejoinder, however, W. C. Myrvold argues that Albert's requirement of separability is too strong and that the question of compatibility therefore remains entirely open. W. C. Myrvold, "Relativistic Quantum Becoming," *British Journal for the Philosophy of Science* 54 (2003): 475–500.

25. J. S. Bell, "On the Einstein, Rosen, Podolsky Paradox," *Physics* 1 (1964): 195. For related discussion see: Niels Bohr, "Discussion with Einstein on Epistemological Problems in Atomic Physics, in *Albert Einstein, Philosopher Scientist,* ed. Paul A. Schilpp (New York: Harper and Row, 1949), 201–241. Also, in the same volume, H. Margenau, "Einstein's Conception of Reality," 243–69, and Phillip Frank, "Einstein, Mach, and Logical Positivism," 269–86. Also, see, J. S. Bell, *Speakable and Unspeakable in Quantum Mechanics* (Cambridge: Cambridge University Press, 1987).

26. Spin components are noncommuting variables.

27. For a highly accessible and informative account and derivation of the Bell inequality, see, Bernard d'Espagnat, "The Quantum Theory and Reality," *Scientific American* 241, no. 5 (1979). The following is a simplified version of the derivation:

Assume first (contrary to QT) that any particle can either have or not have any of the three components and let $N(x^1y^0)$ be the number of single particles having x but not y. Then as a simple matter of the relations between the sets involved:

$$(1)\quad \{x^1y^0\} = \{x^1y^0z^1\} \cup \{x^1y^0z^0\}$$

So that

$$(2)\quad N(x^1y^0) = N(x^1y^0z^1) + N(x^1y^0z^0)$$

But by set-subset relations,

$$(3)\quad N(x^1z^0) \geq N(x^1y^0z^0)$$
$$(4)\quad N(y^0z^1) \geq N(x^1y^0z^1)$$

Substituting these inequalities in (1) we get

$$(5)^*\quad N(x^1y^0) \leq N(x^1z^0) + N(y^0z^1)$$

This inequality cannot be tested directly because of the quantum theoretic impossibility of assigning two simultaneous components to any one particle. The assignment, however, can be made indirectly by an EPR determination on a singlet state. If one of a pair of particles in the singlet state measures x^1 and the other measures y^1, we know the first particle must be an x^1y^0 particle. We can therefore say that

(6) $N(x^1y^0) \propto n(x^1y^1)$

(7) $N(y^0z^1) \propto n(y^1z^1)$

(8) $N(x^1z^0) \propto n(x^1z^1)$

Where this time $n(x^1y^1)$ is the number of proton *pairs* whose individual members have positive but different components, respectively. These proportionalities follow statistically because for any particle having x^1y^0 there is a corresponding pair of particles having respectively x^1y^1, etc. By symmetry, all constants of proportionality are the same.

Substituting in (5)* gives us

(6) $n(x^1y^1) \leq n(x^1z^1)+n(y^1z^1)$ Bell's inequality.

*Note that this is an inequality between numbers of proton pairs whose members are found to have respectively positive but different spin components.

28. Tittel, W., et al, "Violation of Bell Inequalities by Photons More than 10 km Apart," *Physical Review Letters* 81 (1998): 3563–66. Also, "Aspect, A. et al.," *Physical Review Letters* 47 (1981): 460; *Physical Review Letters* 49, no. 91 (1982): 1804.

29. The good Bishop Berkeley knew this, and resorted to making God (or his ideas) the ultimate reality (substance?) behind what humans experience.

30. In addition to the usual Bohm-type experiment in which two particles that were "co-produced" fly off in opposite directions and retain their coherence, some experiments use two independent laser beams to produce the entanglement.

Chapter 4. Interpretation

1. Hillary Putnam, in his very readable synoptic account of attempts at interpreting QT, expresses the problem of interpretation with a question that starts in a similar vein. He asks: "What is it about physical systems that makes them lend themselves to representation by systems of waves?" These are waves whose "amplitudes are complex and not real" and for which "the space involved is a mathematical abstraction and not ordinary space." Hillary Putnam, 134.

2. It is important to note that we are not claiming anything here about what scientists ought to believe or about any essential nature of scientific knowledge. We are saying something about pervasive behaviorisms and underlying beliefs of scientists as strongly evidenced by what they do—namely, how they informally report what they know, how they go about testing that knowledge, and, often, how they go about striving toward new discoveries.

3. In the widely varying contexts of discussion on hidden variables, it is not always clear as to whether "hidden" means "unobservable by current means" or "hidden from the purview of quantum mechanics," that is, "additional to the variables occurring in quantum mechanical description."

4. D. Bohm, *Causality and Chance in Modern Physics* (London: Routledge and Kegan Paul, 1957), 80.

5. Ibid., 91.

6. Ibid., 95. For more discussion see, "A Suggested Interpretation of Quantum Theory in Terms of Hidden Variables," *Physical Review* 85 (1952): 166–79. Also interesting are some comments by H. Reichenbach in "Foundations of Quantum Mechanics" in *Hans Reichenbach, Selected Writings,* ed. R. S. Cohen and M. Reichenbach, (Dordrecht and Boston: Reidel, 1978), 253–54.

7. Hidden variable theorists ultimately assume the classical non-disturbance model according to which any disturbance of the subject matter in the act of measurement can, in principle, be either made negligible or else corrected for, thus denying the irreducibility of uncertainty in quantum measurement.

8. Born's position was somewhat problematic and enmeshed in an issue that roiled the foremost architects of quantum theory in the mid-1920s. I refer here to the problem of what to make (physically) of the "waves" of the quantum formalism. This was a matter of interpretation—that is, one of deciding whether Schrödinger waves represent real physical waves or something else.

To deal with the awkward discontinuity in von Neumann's wave collapse theory, what one might call *Born's original interpretation* takes the Schrödinger wave to be a representation *not* of the actual physical system but of our *knowledge* of the system. This makes the collapse a discontinuity not of the physical system but of our knowledge of the physical system—a "collapse," as it were, of our ignorance rather than of a physical wave. Elementary particles of the actual physical system are thus left as point masses "unreached" and therefore undisturbed by measurement, each having a definite position and momentum at every instant.

In including Born among the hidden variable theorists, I am referring to this phase of his views. At the same time I am regarding as a hidden variable theory any interpretation that allows an elementary particle a definite simultaneous momentum and position despite the fact that we can know *one* but never *both* simultaneously. As is well known, however, that other part of the Born interpretation usually referred to as the *Born probability interpretation* eventually became a standard part of the Copenhagen interpretation, which, as noted earlier, was radically opposed to hidden variable theories—especially those of the kind famously and elaborately formulated by David Bohm. I believe the present characterization of hidden variable theories to be consistent with the literature and I am not alone on the matter. H. Putnam, for example, to whom I am indebted for alerting me to the hidden variable aspects of the "original Born interpretation," explicitly includes Born (as per his original interpretation) among the hidden variable theorists. See, Hilary Putnam, 134–39.

9. In addition to the aim of restoring causality and determinism in the quantum-theoretic account of the microworld, hidden variable theory also envisions the closely related but broader ideal of incorporating both microscopic and macroscopic entities under one unifying theory of physical nature.

10. Oddly enough, von Neumann's projection postulate, positing a wave collapse that proves to be disconnected from anything directly experiential and has, by some critics, therefore been characterized as a "hidden variable," is rather standardly accepted by the very Copenhagenist orthodoxy that is so wary of "extraneous metaphysical constructs."

11. Bohm, 112.

12. The quantum potential is posited as a "hidden potential" in order to account for what, according to QT, is a finite probability (confirmed experimentally) that, upon measurement, an electron can be found on the other side of a potential barrier that would be classically insurmountable by the electron. Again, as in the case of virtual pair production, (see ch. 3, Note 15), it is QED, rather than quantum theory alone, that yields a solid causal explanation of tunneling.

13. For a fascinating and dramatically popularized but highly informative and lucid account of the present status of experimental particle physics along with the great search for the hypothetical and purportedly fundamental Higgs field and Higgs particle, see, *The God Particle,* by Nobel laureate Leon Lederman (New York: Dell, 1993). Even philosophers of science and physicists have probably found his historical perspective and qualitative synoptic presentation of a broad subject instructive, as I did.

14. Bohm, 95–96.

15. I first became familiar with early versions of J. M. Jauch's views as one of his graduate students at Princeton. For guidance on some of the historical detail regarding von Naumann's impossibility proof and its repercussions, however, I am indebted to Max Jammer's very accessible and philosophically enlightened discussion in his classic work, *The Philosophy of Quantum Mechanics* (New York: John Wiley and Sons, 1974) 265–78. Also, see his: *The Conceptual Development of Quantum Mechanics* (New York: McGraw-Hill, 1966) for related material in the history of QT.

16. The domain of this calculus is a "lattice" or the set of all "propositions" of a physical system where the term *proposition* refers to a yes-no experiment conceived as an element to which all physical measurements can be demonstrably reduced. In the context of this "physical calculus" the authors then define a set of strategically chosen interpropositional relationships that when appropriately interpreted express the structure of general quantum mechanics. These authors then show that, if any propositional system were to admit hidden variables, then every pair of propositions in the system would be "compatible," and such compatibility is contrary to experience. Consequently, quantum mechanics, which is true to experience, cannot be interpreted in terms of (i.e., cannot admit) hidden variables. The sense of "compatibility" here is one that is analogous to the commutativity of two quantum variables. See J. M. Jauch and C. Piron, *Helvetica Physica* Acta 36 (1963): 827–37.

17. Ibid., 837.

18. J. S. Bell, "On the Problem of Hidden Variables in Quantum Mechanics," *Reviews of Modern Physics* 38 (1966): 447–75.

19. For more recent work in this area see: Jeffrey Bub, op.cit., especially ch. 4, 115–34.

20. D. Bohm and J. Bub, "A Proposed Solution of the Measurement Problem in Quantum Mechanics by a Hidden Variable Theory," *Reviews of Modern Physics* 38 (1966): 453–69.

Bohm's co-author, J. Bub, offers a brief discussion of this hidden variable approach particularly as it relates to philosophic issues about quantum measurement and the ontological relationship between what is being measured and what is doing the measuring. See J. Bub, "Hidden Variables and the Copenhagen Interpretation—A Reconciliation," *British Journal for the Philosophy of Science* 19 (1968): 185–210.

21. The price of such a move, for any fundamental law of nature, is a blistering loss of mathematical and computational manageability. The present linear formalism of QT is computationally daunting enough. Indeed, the cases for which the Schrödinger equation can be solved exactly are very few (though very important). These are the very special cases in which the potential energy is a simple function of position as, for example, in the case of the hydrogen atom, the linear oscillator, and the particle that has one constant potential energy value for one range of position, another value for another range, etc. For all other cases, solving the Schrödinger equation, *even in one* dimension (i.e., as an ordinary differential equation), can be a complicated affair requiring the use of variational, perturbative, or other methods of approximation.

22. There is no way, it seems, of overestimating the importance of thought experiments at any stage of theory development. Examples of this form of creative thought are legion in the history of science, possibly suggesting that thought experiments may have played at least as great a role as real ones. For two famous examples, we recall Galileo as probably never having dropped anything from the Leaning Tower. He could easily have imagined a large stone to be merely several smaller ones tightly bound together with the stone-to-stone proximities in no way affecting their acceleration. Likewise, Einstein imagined himself hurtling forward possibly at the speed of light while projecting a light beam straight ahead yet unable to intuit anything other than that the light beam would still advance ahead of him at an unchanged relative speed.

The remarkable fruitfulness of thought experiments in theorizing, if viewed uncritically, can easily arouse epistemological wonderment or even draw the charge of a priorism. This extremely powerful technique of theorists from Galileo to Einstein (and all in between), however, is no less transparent than any other hypothetical (what-if) thinking. In such conjectural thinking, background commonsense "know-how"—usually as solidly established knowledge but often as little more than hazy common sense impressions or "intuitions"—is brought to bear on an imagined experiment in order to come up with a plausible imagined outcome.

23. Notably, Jeffrey Bub in his book, *Interpreting the Quantum World* (Cambridge: Cambridge University Press, 1997).

24. The principle can, for example, be applied to require that the expectation values of the dynamical variables calculable for a quantum wave packet satisfy the classical laws of motion, provided the wave packet's peak is fairly sharp and the momentum of the packet is as precise as the uncertainty principle allows. Such a requirement leads to satisfactory definitions of

the expectation values of dynamical variables such as energy velocity and momentum.

25. Once the wave aspects of material particles were established empirically, Bohr introduced complementarity, an idea closely linked to his correspondence principle. Bohr's complementarity takes wave and particle manifestations to be "complementary" aspects of one and the same reality, so that complete physical descriptions of phenomena must include both aspects. What is more, the aspects of undularity, stability, and corpuscularity of phenomena are, together, strongly suggestive of standing waves under appropriate boundary conditions. Thus, any particle, say, an electron, moving freely and classically may, with approximation, be regarded as moving classically around a central force at a relatively very large distance from the center. Imposing the Bohr-Sommerfeld quantum condition of standing waves (under the boundary condition of a circular orbit) gives us: $\int_0 pds = nh$. Given the large dimensions involved and the ultra smallness of h, n in this case must be hugely large. But Bohr's correspondence principle requires that this quantum equation—established via complementarity—for the *classical* limiting case ($n \rightarrow \infty$) applies also to the non-classical case in which n is small. (Small values of n, of course, take us to the quantum mechanical domain of very small dimensions comparable in size to h.)

Chapter 5. Fresh Starts

1. This idea was first proposed by Hugh Everett, "Relative State Formulation of Quantum Mechanics," *Reviews of Modern Physics* 29 (1957): 454–62. For a readable exposition of Everett's ideas, see. B. S. DeWitt and N. Graham, *The Many Worlds Interpretation of Quantum Mechanics* (Princeton: Princeton University Press, 1973). Also, B. S. DeWitt, "Quantum Mechanics and Reality," *Physics Today* (1970): 30–35. For an evaluative response, see, J. A. Wheeler, "Assessment of Everett's Relative State Formulation of Quantum Theory," *Reviews of Modern Physics* 29 (1957): 463–65.

2. As M. Lockwood would have it, a realist interpretation of quantum mechanics does not imply a commitment to non-local interactions of some kind. "One can avoid this implication by rejecting the assumption that when a measurement is carried out, *one* of the possible outcomes occurs *to the exclusion* of all the others." M. Lockwood, "'Many Minds' Interpretation of Quantum Mechanics," *British Journal for the Philosophy of Science* 47 (June 1996): 164.

3. As far as I know, the unlikely attempts to derive quantum probabilities from the core formalism of QT have been unsurprisingly few and uniformly unsuccessful. Of these, a mention of two due to B. S. De Witt and H. Everett can be found in Lockwood, 173–74.

4. See: H. Everett, "Relative State Formulation of Quantum Mechanics," *Reviews of Modern Physics* 29 (1957): 454-62; also J. Wheeler and H. Zureck, eds., *Quantum Theory and Measurement* (Princeton: Princeton University Press, 1987).

5. More precisely, if a conscious state of the observer and the cosmic remainder are represented by the vectors, $|C\rangle$ and $|R\rangle$, respectively, in Hilbert space, then the tensor product, $|C\rangle\,|R\rangle$, would represent a current world. One could of course define many such worlds with respect to tensor products of all conscious states in the entire universe.

6. D. Deutsch, "Quantum theory as a Universal Physical Theory," *International Journal of Theoretical Physics* 24 (1985): 1–41.

7. Lockwood, 172.

8. I use the term *diagnostically* here for the kind of inductive reasoning sometimes referred to as "hypothesizing," to mean formulating an assumption (hypothesis) for the purpose of explaining some set of findings. For further discussion, see: S. Cannavo, *Think to Win* (Amherst, NY: Prometheus Books, 1998), 204–31.

9. Lockwood, 172.

10. David J. Chalmers, *The Conscious Mind* (Oxford: Oxford University Press, 1996), 352–53.

11. Any vector, **C**, in an n-dimensional vector space can be described (decomposed) as a linear combination of the vectors, $\mathbf{e}_i$, of a set of linearly independent vectors, such that $\mathbf{C} = \Sigma_n c_i e_i$. When the n $\mathbf{e}_i$ span the space, then the $\mathbf{e}_i$ form a basis and is said to "constitute a complete set."

The multiplicity of possible bases is a difficulty that also afflicts collapse theory. The von Neumann projection postulate does not tell us which, of the many possible decompositions of an evolving system, is the one that on the occasion of measurement is projected into one or another component of that decomposition. Nor does the Schrödinger formalism provide such a determination.

To illustrate, consider the often mentioned measurement on a microparticle, m, whose spin, S, along some arbitrary coordinate axis can have only two possible values, either spin-up (S^U) or spin-down (S^D). Then, let one possible state of the detector dial read U and the other D, and let the corresponding observations (states of awareness, states of consciousness, experiences) on the part of the observer be O^U and O^D, respectively. According to QT and using the ket vector notation, we can, on the occasion of a spin measurement in, say, the x direction, represent the state of the composite system, S_c, consisting of spin, dial reading, and observer awareness as: $1/\sqrt{2}\ |S_x^U\rangle|U\rangle|O^U\rangle + 1/\sqrt{2}\ |S_x^D\rangle\ |D\rangle|O^D\rangle)$. But we can also represent S_c as, for example,

$$1/\sqrt{2}\ |S_y^U\rangle\ \{1/\sqrt{2}\ |U\rangle|O^U\rangle + 1/\sqrt{2}\ \rangle|D\rangle|O^D\rangle\} + 1/\sqrt{2}\ |S_y^D\ \rangle\ \{1/\sqrt{2}\ |U\rangle|O^U\rangle - 1/\sqrt{2}\ |D\rangle|O^D\rangle\}$$

where S_y is the spin in the y direction. There is no limit to the number of ways that S_c can be represented, depending on the choice of vector basis. This particular example, however, is especially suited for simplicity of exposition, and I owe it to M. Lockwood. Lockwood, 167.

12. H. D. Zeh, "The Problem of Conscious Observation in Quantum mechanical Description," *Epistemological Letters* 63 (1981).

13. D. Albert and B. Loewer, "Interpreting the Many Worlds Interpretation," *Synthese,* 77 (1988): 195–213.

14. "For the many minds theorist, the *appearance* of there being a preferred basis . . . is to be disregarded as an illusion—an illusion to be explained by appealing to a theory about the way in which *conscious mentality* relates to the physical world as unitary quantum mechanics describes it." Lockwood, 170.

15. Lockwood, 174, ref: D. Albert, B. Loewer, 195–213.

16. As "nonphysical entities," minds would not, themselves, evolve according to physical laws, e.g., the laws of quantum physics. MWI and MMI theorists try to circumvent this limitation by vaguely "associating" conscious states with evolving physical observers.

17. Lockwood, 175.

18. M. Hemmo and Itamar Pitowsky, "Probability and Nonlocality in Many Minds Interpretations of Quantum Mechanics," *British Journal for the Philosophy of Science* 54 (2003): 228.

19. Ibid., 228.

20. For details, see: Ibid., 238–43.

21. The supervenience principle embraced by some current philosophers of mind, requires that minds be "supervenient on" physical structures such as brains—*supervenience* being a very general type of dependence relation that has come to play a central role in current mind-body philosophy. To say that mental states are supervenient on brain states is to say that there are no two "possible" situations in which the mental states are identical (in the sense of indiscernibleness) and the brain states different. For further explication of this notion as regards various senses of "possible" in this definition, see: Chalmers, 32–41.

Full conformity with the supervenience principle requires more than an avoidance of the mindless hulk predicament. It must also disallow "free reigning" nonphysical minds that can, at random and regardless of their state of consciousness, "hop on" to any brain regardless of its physical structure.

22. Hemmo and Pitowsky, 235–38.

23. More formally, we refer to the question of which term of the evolving state vector, $|\Psi\rangle$, an individual mind will randomly "choose" as new superpositions arise.

24. Chalmers, 354.

25. Lockwood, 177.

26. Ibid., 178.

27. More formally, Lockwood assumes that there is a mixed state of mutually orthogonal pure states that comprise a *basis* (consciousness basis) for his multimind. A maximal experience (a conscious event) is assumed to correspond to some pure state $|\varphi\rangle$ belonging to this basis set. For a detailed account see: Lockwood, 176–82.

28. Ibid., 178–86.

29. Ibid., 173.

30. Ibid., 172–73.

31. "Still it is a fair question why consciousness and perception should favor the states that they do." Lockwood, 185.

32. For an excellent introduction to Decoherence theory, see, Guido Bacciagaluppi, "The Role of Decoherence in Quantum Theory," *Stamford Encyclopedia of Philosophy* (Winter 2003), ed. E. N. Zalta. I am indebted to this very lucid, informal account for some of its philosophical insights and summative aspects.

33. H. D. Zeh, "Basic Concepts and their Interpretation," Section 2.5, in E. Joos et al., *Decoherence and the Appearance of a Classical World in Quantum theory* (Heidelberg: Springer-Verlag, 2003).

34. It is upon such entanglement that the systems, even after eventual separation, can no longer be described as before, i.e., as strictly and separately independent states to be regarded as no more than an account of what an observer has done to the "system" and of what she/he has subsequently observed in the experiment.

35. The total quantum state of the composite system Ψ_c is represented as the tensor product $|\Psi(x_i,\ldots,x_n)\rangle \otimes |E\rangle$, the first factor being the state of the original system and the second the state of the environment. More explicitly, this product is $|\Psi_c\rangle = \Sigma_i \alpha_i |\psi_i\rangle \otimes |\varphi_i\rangle$ where $|\psi_i\rangle$ are component states of the original system and $|\varphi_i\rangle$ alternative eigenstates of the pointer positions (in this case, the environment). The α_i are complex coefficients and $|\psi_i\rangle \otimes |\varphi_i\rangle$ are product states.

36. G. C. Wick, A. S. Wightman, and E. P. Wigner, "The Intrinsic Parity of Elementary Particles," *Physical Review* 88, no. 36 (1952): 101–105. See also E. Joos et al., *Decoherence and the Appearance of a Classical World in Quantum Theory* (Heidelberg: Springer Verlag, 2003) ch.3.

37. See, A. J. Leggett, "Schrödinger's Cat and Her Laboratory Cousins," *Contemporary Physics* 25: 583–94. For this reference, I am indebted to G. Bacciagaluppi (2003).

38. For a formal account, see: W. H. Zurek, "Decoherence, Einselection, and the Quantum Origins of the Classical," *Review of Modern Physics* 75 (July 2003).

39. J. Bub has argued that even granting DT "the luxury of a preferred bases, it fails under consideration of an observer's beliefs about the outcome of some experiment (what the observer believes, believes she believes, etc.)." That is, it explains what the observer believes an outcome to be apart from what that outcome actually is. See: J. Bub (1997), ch. 8.

40. To explain why, in measurement, we observe one definite (discrete) outcome, and nothing else, rather than a superposition of different ones, the practice, in various quantum-theoretical contexts (e.g., quantum field theory) has been to postulate superselection rules. Though some theorists have sought to implement DT for the justification of such rules, a theoretical basis for them would require a modification (generalization) of the Schrödinger formalism so as to allow some variables that commute with *all* observables. Absent such a modification, superselection rules remain essentially ad hoc supplementations in QT. See A. S. Wightman, "Superselection Rules; Old and New," *Il Nuovo Cimento* 110B (1995): 751–69.

41. For a more detailed and formal account See: H. D. Zurek, "Basic Concepts, and Their Interpretation," in *Decoherence and the Appearance of a Classical World in Quantum Theory,* ed. D. Giulini et al. (Berlin: Springer, 1996), ch. 2, 16. Also, "Decoherence, Einselection, and the Existential Interpretation," *Philosophical Transactions of the Royal Society of London* A 356 (1998): 1793–1820.

42. See, for example, P. Pearle, "Reduction of a State Vector by a Nonlinear Schrodinger Equation," *Physical Review* D13 (1979).

43. This proposal was made by G. C. Ghirardi, A. Rimini, and T. Weber in a much-discussed paper (1985) and has become known as the GRW interpretation. See: G. C. Ghirardi, A. Rimini, and T. Weber, "A Model for a Unified Quantum Description of Macroscopic and Microscopic Systems," in *Quantum Probability and Applications,* ed. L. Acardi et al. (Berlin: Springer, 1985). Also by the same authors, see: "Unified Dynamics for Microscopic and Macroscopic Systems," *Physical Review* D34 (1986): 470. For my account of GRW theory, I found particularly helpful, the broad and highly enlightening discussion of collapse theories by Giancarlo Ghirardi himself, in "Collapse Theories," *Stanford Encyclopedia of Philosophy* (January 28, 2007).

44. A statistical distribution whose frequency function is of the form $f(x) = m^x e^{-m}/x!$ for $x = 0,1,2,...$, where m is called the mean or variance of the distribution. (The two are equal for Poisson distributions.) The distribution tends to occur for events that are very improbable but occur occasionally because of the large number of trials involved.

45. Localizations occur by chance at any time. As soon as a hitting occurs for an ith-particle at some point, x, in a given system of n distinct particles, the wave function, $\Psi(q_1 ...q_n)$, for that system is instantly multiplied by an appropriately normalized Gaussian distribution function, $G(q_i,x) = C \exp[\{-1/(2d2\}(q_i-x)]$, thus resulting in a new wave function, $\Psi_i(q_1 ...q_n ; x) = \Psi(q_1 ...q_n) G (q_i,x)$. The quantity, d, representing the "narrowness" of G is a measure of the localization accuracy, while the constant, C, is assigned a value for which the integral of the probability density, P(x), over the entire space comes out equal to 1.

The function, Ψ_i, provides the means for determining where the localization (hitting) takes place. This results from the assumption that the probability that a localization taking place at x is the integral of $|\Psi_i|^2$ over the 3n-dimensional space. This assumption assures that the desired probability is larger wherever the probability of finding the particle at x in the standard quantum mechanical description is larger. The constant, C, is, of course, so chosen that the integral of the probability density over the whole space is 1. Finally, GRS assumes that hittings occur at random according to a Poisson distribution and with f as the average frequency.

46. For this reason, other versions of wave collapse theory have attempted to link wave collapse to some physical threshold—as, for example, gravitational forces in relativistic space-time. See, Roger Penrose, *Shadows of the Mind* (Oxford: Oxford University Press, 1994).

47. In actual practice, such interpretation is not usually established formally. The descriptive (i.e., nonlogical) terms of physical theories are usually tied to experimental ideas not by explicit definitions but by meaning conventions generally understood informally by those who use the theories.

48. The nature of *scientific explanation* is a controversial issue in the philosophy of science. It is a problem that has spilled over into even deeper controversy over the nature of scientific theory. These closely related issues have been thrashed over a long trail from early–twentieth-century positivism through logical empiricism (C. G. Hempel), scientific realism (W. Salmon), and societal relativism (T. S. Kuhn).

49. The term *quantum logic* for what we are calling "formal interpretations" of QT is used, for example, by M. Jammer (1974), 341–416.

Chapter 6. Explanatory and Algorithmic Nomologicals

1. A live discussion at which I was present as a graduate student. Fine Hall, Princeton University, *circa* 1945.

2. There are scientific laws that seem to have no causal content but have limited unifying power and definite descriptive content that is clearly existential and systematically central enough to deny them algorithmic status. Examples are laws such as those expressing the invariance of the speed of light in free space for all observers or those expressing the values of the physical constants of nature (the gravitational constant, Planck's constant, etc.).

3. Dynamical spaces are *real* spaces not only of the Cartesian space-time sort but also of more abstract sorts such as "phase spaces" with momentum and position as coordinates or higher-dimensional (though purportedly "real") spaces such as those of spin theory.

4. See, C. G. Hempel and P. Oppenheim, "Studies in the Logic of Explanation," *Philososphy of Science* 15 (1948): 135–78.

5. S. Cannavo, *Nomic Inference* (An Introduction to the Logic of Scientific Inquiry) (The Hague: Martinus Nijhoff, 1974), 125–46. For an excellent account, which I found helpful as a historical sketch on scientific explanation, see Philip S. Kitcher, "Philosophy of Science," *Britannica Online Encyclopedia*, www.britannica.com/eb/print?articleleId=108542&fullArticle=true&tocId=271817.

6. For two highly influential works arguing that tracing causes is essential for scientific explanation, see: Wesley Salmon, *The Foundations of Scientific Inference* (Pittsburgh: University of Pittsburgh Press, 1964) and also his *Scientific Explanation and the Causal Structure of the World* (Princeton: Princeton University Press, 1984). For additional discussion, see: S. Cannavo (1974), 125–46.

7. While—as in the Humean account—constant conjunction remains at the core of the necessity in causal relations, it is not enough for distinguishing "cause" from accidental concomitance. I would submit that a more complete

analysis must couch the concomitance in the context of a quantitatively detailed theory with appropriate ontological content—content that usually includes such quantities as motions, forces, masses, fields, etc. For example, on a given point mass, the constant conjunction of a net force (computed by vector addition) and an acceleration in the same direction as the force takes on the aura of causal necessity when we express the relation algebraically and precisely as a direct proportionality between the force and the acceleration, that is, when we express the relation as Newton's second law of motion. For a discussion of other characteristics of physical laws expressing necessity see: S. Cannavo (1974), 195–286.

8. The physical interactive relationships that characterize the dynamical subject matter of causally explanatory theory are typically represented by mathematical equations expressing functional space-time relationships (field equations, wave-type equations, etc.) that formally mirror the kind of causal relations involved. Examples of such causal theory are classical dynamics, electromagnetism, kinetic theory, molecular physical chemistry, astrophysics, relativity, and so on to include nearly all basic classical theories along with the quantum field theories and string theory.

9. Most certainly, it is philosophically important to give full and due recognition to the systemizing effect of all non-algorithmic scientific theory. In the interest of perspicuity, however, I believe it desirable to construe *explanation* either in causal terms or in terms of strong theoretical unification. Simple (weaker) unification under covering laws can then be expressed with the weaker term, "indicate." Accordingly, a positive result on a test together with the observational laws connecting such a result with some disease would be said to *indicate* (rather than *explain*) the presence of the disease. Fortunately, however, for our purposes here we need make no strong commitment on distinctions of this sort.

10. For a discussion of hypothesizing as diagnostic inference see, Cannavo (1998).

11. The radiation law in point was the Wien displacement law, and h was what became known as Planck's constant.

12. As it turned out on the basis of the developing formalism, two noncommuting variables, p and x, are incompatible with respect to simultaneous measurement in the sense expressed by the uncertainty principle. Demonstrably, a measure of their incompatibility is given by the commutator: $s[p_x,x] = px - xp = \Sigma_i \{\partial p_i/\partial x_i \, \partial x_i/\partial p_i - \partial x_i/\partial p_i \partial p_i/\partial x_i\}$.

13. The matrix representation of physical quantities can also be introduced into the quantum mechanical formalism on the basis of formalisms that came a bit later. These were the time dependent Schrödinger equation and Born's statistical postulate as follows: Start with the general solution, Ψ, of the Schrödinger equation, expressed as a summation (superposition) over i sinusoidal oscillations. Then write the expression for the probability density, $\Psi\Psi^*$, as a double summation over i and j such oscillations. This probability result can then be used to write down the weighted mean for any physical quantity, F, (e.g., the position or momentum) that is a function

of the coordinates. The resulting expression (a double summation) contains quantities, Fij, that form a two-dimensional array or matrix which, moreover, is easily shown to have the analytically important and highly useful computational property of being Hermitian, i.e., interchanging the order of the subscripts of any nondiagonal element converts it into its complex conjugate.

14. There are four quantum numbers (n, l, m_l, and m_s) that define the *quantizations* that give quantum mechanics its name, and uniquely specify the quantum state (energy level) of an electron in an atom. Three of these are implied by the Schrödinger formalism, which leads to equations whose solutions—if they are to be useful—require them along with their numerical restrictions. The fourth quantum number results from the introduction of electronic spin. The derivability of the Pauli Principle and of the quantum numbers from the quantum formalism cannot, however, invest them with explanatory content if the formalism itself has none. So while these results suffice to *generate* the periodic table they do not *explain* it. For the desired explanations we must go to quantum electrodynamics and quantum field theory.

15. From the viewpoint of interpreting QT, Schrödinger's wave formalism is more engaging than Heisenberg's matrix mechanics largely because of the ontologically suggestive wave equation. In actual practice, however, QT as presently used is in fact a mixture of the two formalisms (but remains, of course, no less algorithmic for it). The use of matrix algebra is especially convenient in QT when the number of possible values a physical quantity can have is finite. An example, of such a quantity is spin, whose projection along any fixed axis can have only two values, + or –.

16. De Broglie's wave hypothesis was indeed a daring "symmetrizing" move and, as it seemed, an ontologically promising one. Like other such moves, it triggered theory creation—in this case, the formulation of wave mechanics. As the start of an explanatory theory, however, it proved to be a false dawn. For, apart from the Born probability interpretation, matter waves remained, as we have seen, resistant to any sort of clearly physical interpretation.

17. Some writers attribute a different turn of mind to Schrödinger. Straightline geometric optics does not correctly predict what happens when dimensions in an optical system get close (in smallness) to the wavelength of the radiation. In such a case, one goes to physical optics. Similarly, straight line Newtonian mechanics does not correctly say what happens when the dimensions of a physical system approach those of the De Broglie wave. For such a case, Schrödinger might have gone from the straight line physics of Newtonian mechanics to the wave physics of physical optics and started with its basic wave equation, namely, $\partial^2 f/\partial x^2 + \partial^2 f/\partial y^2 + \partial^2 f/\partial z^2 = 1/u^2\ \partial^2 f/\partial t^2$. To get to the wave equation for Ψ, he could then have proceeded by letting $f = \Psi e^{2\pi ivt}$ so as to focus only on periodic solutions. This, in turn, would lead to the quantizations that give quantum mechanics its name. But again, the highly prescriptive and algorithmic approach of the overall strategy is obvious.

18. The mathematical analysis for solving the time-dependent Schrödinger equation yields the time-independent. form of the equation. This form of the Schrödinger equation is what finally leads to the quantizations

that gives quantum mechanics its name. That is, it turns out that meaningful solutions can be obtained only if the energy, E, is constrained to discrete values, called *eigenvalues*—the solutions that correspond to these eigenvalues being called *eigenfunctions.* By "complex" we mean, "having both a real and an imaginary part."

19. For an introductory but fairly technical account of Dirac transformation theory, see,"Dirac Equation," *Wikipedia* (September, 14, 2007) http://en.wikipedia.org/wiki/Dirac_equation.

20. It features Hermitian matrices, and any Hermitian matrix representing a physical quantity can always be diagonalized in terms of a unitary transformation matrix. In the resulting representation, the wave functions are eigenfunctions and the measurable quantities are the eigenvalues of the corresponding operators. These eigenvalues are real numbers and are therefore directly relatable to physical measurement and confirmation.

21. The Liouville equation is: $\partial\rho/\partial p_i + \Sigma_n[\partial H/\partial\rho \ \partial\rho/\partial x_i - \partial H/\partial x_i \ \partial\rho/\partial p_i] \equiv \partial\rho/\partial t + [\rho, H]$ where ρ is the distribution function in a multidimensional phase space whose coordinates are the positions x_i and the momenta, $p_{i\,;\,.}$ H is the Hamiltonian; and the expression, [ρ, H] is called the Poisson bracket.

22. These are the abstract symmetries that are featured in group theory.

23. A familiar and more "symmetrized" version of this analogy is: Just as particles (photons) may be associated with electromagnetic waves, waves (matter waves) may be associated with matter particles.

24. The elaboration of Hamiltonian theory so as to provide a general procedure for the task of solving mechanical problems (involving the motion of a system of particles), leads to the Hamilton-Jacobi equation whose solution, known as *Hamilton's principle function* makes it possible to carry out the task. The mathematical behavior, in phase space, of Hamilton's principle function is remarkably suggestive of wave propagation. The function defines "surfaces" that propagate in the same manner as wave surfaces of constant phase. The association of waves with Hamilton's principle function, however, is more than casual or superficial. The associated waves can be characterized with determinable frequencies, amplitudes, and velocity. Indeed, with close approximation, the isomorphism carries right into geometric optics. For a lucid mathematical discussion of this and the historical bearing it may have on the discovery of wave (quantum) mechanics see: H. Goldstein, *Classical Mechanics* (Reading, MA: Addison-Wesley, 1959), 273–314.

25. As the velocity, v, of a body approaches the velocity of light, its mass increases without limit (special relativity). The gain in its energy due to the motion is its kinetic energy, KE. Thus, in terms of mass-energy equivalence, $KE = mc^2 - m_0c^2$. As we approach the velocity of light, the momentum, mv, approaches mc and the kinetic energy approaches mc^2. For a photon the rest mass is zero and so the kinetic energy, mc^2, is equal to the total energy, E, where $E = h\nu = mc^2$. Consequently, $p = h\nu/c$, and $p = h/\lambda$.

26. The type of quantization that Bohr postulated for the orbits of the hydrogen atom finds its general quantum mechanical expression in

what is known as the Bohr-Sommerfeld quantum condition. Originally it was formulated by Sommerfeld, who gave it a generalized mathematical form that allowed the atom only a discrete set of stationary (energy) states. Sommerfeld's formulation is most simply stated for a particle moving classically in one-dimensional periodic motion. He assumed that the integral with respect to the coordinate, x, taken over the varying momentum, p (i.e., the phase integral), must be an integral multiple of Planck's constant. That is, $\int_0 pdx = nh$ where p is the particle's momentum, n is an integer, and h is Planck's constant. This requirement (applicable to periodic motion, and therefore to the quantum states represented by the Ψ-function) threads, in one form or other, through all of QT. With the advent of wave mechanics and its application to the case of a quantum particle in a force field that changes with position, this quantum condition resurfaces more formally. For such a case, we seek periodic solutions (Ψ) of the time-independent Schrödinger equation—solutions that represent standing waves under appropriate boundary conditions (analogously to the case of classical physical waves).

Carrying through the analogy with familiar physical waves, we require that Ψ and its space derivatives be continuous, finite, and single valued throughout configuration space. We also allow for a possible probability interpretation, so that the function, Ψ must be normalized, i.e., $\iiint |\Psi|^2\, dxdydz = 1$ over all possible values of x, y, and z. This last condition delivers the desired quantization. It can be met only for certain *discrete values* (eigenvalues) of the energy, and it is these that are associated with "stationary states" of the system, that is, with states for which the probability assignable to any observable does not vary with time. Such states correspond to different eigenfunctions or solutions of the *time-independent* Schrödinger equation.

The quantum condition is similarly imposed on solutions of the more general Schrödinger equation involving time. In this case, solutions are obtained by the method of separation of variables. Of the resulting two equations, one involves the time and yields a solution that tells us that whatever the wave function, Ψ, turns out to be it is going to be sinusoidally time dependent. This dependence turns out to have the form of a complex exponential whose real part is not a solution of the Schrödinger equation, thereby establishing the inherently complex nature of Ψ as a quantity. And this of course poses a stiff challenge in any attempt at interpretation. The other equation resulting from the analysis is none other than the Schrödinger equation without the time, for which, again, there are acceptable solutions *only if the energy is restricted to discrete levels* (stationary states). Very generally, the quantum condition in QT represents an ontological move toward explanation, in that the condition is tantamount to postulating the *existence* of a quantum of action.

27. Serendipitously, the set of symmetry operations involved in any of the gauge symmetries, constitutes a *group*—i.e., a special type of set that embodies symmetry or invariance under certain operations. Group theory, therefore, which, as pure mathematics, antedates QFT by more than a century and was once thought to be totally removed from anything physical, is the ready-made tool for dealing with abstract symmetries. In particular, QFT calls

for the theory of continuous or *Lie groups* which have the analytically useful property that the functions relating their elements are differentiable and therefore mathematically and conveniently "well-behaved."

28. A scalar field assigns to each point in a space a scalar (a quantity with no directional properties). By contrast, a vector field assigns to each point in a space a vector (a quantity with direction). The Higgs boson, associated with the Higgs field, has zero spin. It therefore has no angular momentum (a directional quantity or vector). Moreover, it functions to assign mass to entities at every point, and mass is a scalar. As a scalar field, the Higgs field therefore differs in a significantly formal manner from the force fields (vector fields) of QFT.

Chapter 7. A Modest Proposal

1. For a similar observation regarding the quantum formalism, see: N. Maxwell, "A New Look at the Quantum Mechanical Problem of Measurement," *American Journal of Physics* 40 (1972): 1431. For bringing my attention to this particular reference, I am indebted to Max Jammer (1974), 520.

2. See: J. Bub, (1997) esp. ch. 4, 115–34.

3. Ibid., 238–39.

4. J. M. Jauch and C. Piron, "On the Structure of Quantal Proposition Systems" *Helvetica Physica* Acta 42 (1969): 842–48. Also, "Can Hidden Variables Be Excluded from Quantum Mechanics?" *Helvetica Physica* Acta 36 (1963): 827–37.

5. We might recall here how, under the pressures of the measurement problem, the most recent attempts to assign full dynamical content to the quantum-theoretic formalism have led to MWI and MMI. We have also noted, however, that these interpretations raise hugely difficult metaphysical and epistemological issues that cast serious doubt on their confirmability, and consequently on their *plausibility,* if not their very intelligibility.

6. As described, for example, by the standard model.

7. Observational possibilities open up only when there is a "breaking of symmetry" which gives rise to the various interactions and the resulting "gluonic" particles, which, though presently unobservable, may possibly be observed in the future.

8. The primordial soup, cosmological theory tells us, was formed a mini-micro-instant after the big bang in a symmetry-breaking differentiation (from the first homogeneous blob) into quarks and gluons. Physicists working with colliders on both sides of the Atlantic have very recently detected what may be signs of this highly dense material.

9. At high temperatures, inconsistencies arise in the standard model. When the Higgs field is introduced, however, the inconsistencies vanish even at energies of 10^{12} electron volts. This hypothetical entity of gauge theory may also serve to account for the existence of rest mass in the universe. The

explanation of how mass is generated in a particle may be given in terms of how closely the particle is coupled to the Higgs field.

10. As hidden variables par excellence, gluinos cannot be themselves detected; they break down too quickly. The products of their decay, however, are thought to be elementary sparticles, i.e., lightest supersymmetric particles (L.S.P.s) whose existence can be inferred by the detectable missing energy they carry away. Conclusive failure to detect any such signs of gluinos with the available energy at CERN would cast serious doubt on their existence and could put supersymmetric field theory in rather serious jeopardy.

11. For this implausible claim, see a journalist's argument that the most important scientific discoveries are behind us. J. Hogan, *The End of Science* (New York: Broadway Books, 1996).

Chapter 8. Quantumization

1. Virtual particles in the churning "quantum foam" survive transiently for only ultra-brief moments of the order of 10^{-43} seconds.

2. We might recall, here, that the appropriateness of this allowance stems not only theoretically from QT, it is also mandated by nature itself, given the results of the Bell inequality experiments of the past several decades.

3. In my discussion of the quantumized theories—quantum electrodynamics, quantum field theory, and string theory—I am deeply indebted to the physicist Brian Greene, who traces the conceptual and historical development of these theories (especially string theory) in his *The Elegant Universe* (New York: W. W. Norton, 1999), 380–82. Written primarily for a general audience, this much acclaimed, well-illustrated account can illuminate the intuition of even the more specialized reader.

4. According to special relativity, moving a particle suddenly cannot result in any *instantaneous* change (energy transfer) on a neighboring particle. A change can and will occur only after a finite time interval. But if energy is to be conserved at all instances, where is the energy during the time interval? An answer lies in positing the existence of a field to fill the gap and thus make the conservation of energy possible. The application of quantum mechanics, however, leads to a quantization of the field so as to yield messenger particles (photons). Particle interactions via the transmission of the electromagnetic force are then explained as "exchanges" of photons.

5. Reflection easily shows that any commonsense analogy of the billiard ball type cannot quite describe the manner in which messenger particles mediate force between electrically charged matter. While a mechanistic intuition of this sort might do for repulsion as between like charges, it fails for attraction as between unlike charges.

6. The quantization of a field results from a basic postulate of quantum field theory drawn from the quantum mechanical context This is the assumption that at any point in a space, the intensity of the field at that

point yields the probability of finding an "appropriate" particle at that point, i.e., a particle of the sort associated with that field. Intuitively, this may be seen as a "segmentation" or "quantization" of the field.

7. Not all physical theories are renormalizable, and fortunately not all of them need it. But theoreticians, reluctant to inhibit their engagement with a developing theory, see renormalizability as a highly facilitating and therefore desirable feature in the construction stages of theories.

8. As an explanatory, dynamical framework, gauge theory assigns to fundamental particle types appropriate quantities of *strong* and *weak charge* that determine how each type is influenced by the strong and weak forces. For example the assignment of a fanciful "color charge" to quarks and the imposition of invariance to certain shifting color changes (a gauge symmetry) implies the existence of the strong force. The symmetry involved here, generally referred to as a "gauge symmetry," is an abstract symmetry acting on a complex space. The color shifts are transformations acting on the "force coordinates" of a quark, in this case, the "color coordinates."

9. In this discussion, I am indebted to the late physicist Heinz R. Pagels, whose *The Cosmic Code* (New York: Simon and Schuster, 1982) offers a remarkably illustrated and lucid *qualitative* discussion of gauge field theory, highlights many of its details, and provides valuable insights especially regarding the role of symmetry in the theory. I also profited from his comments on realism in contemporary physical theory, at the time of his participation in the Ethyl Wolfe Humanities Seminar at Brooklyn College in the early eighties.

10. Depending on certain formal symmetry properties of their corresponding quantum mechanical operators, all elementary particles fall into two types: those that, in an ensemble of identical particles, obey Bose-Einstein distribution statistics and those that obey Fermi-Dirac statistics. Accordingly they are classified either as *bosons* or *fermions*, respectively. Theoretical analysis in QFT reveals a fundamental relation between the statistics obeyed by a particle and quantum mechanical state transformations under spatial rotations. This relation in turn leads to the conclusion that all particles with integral spin are bosons and all those with half-integral spins are fermions.

QFT is a theory of force fields that have been quantized to yield messenger exchange particles. The quantized electromagnetic force is a photon, the quantized strong force is a gluon, and the quantized weak force is a weak gauge boson.

11. Supersymmetry greatly expands the ontology and explanatory force of QFT by requiring that every elementary particle has a superpartner whose spin differs from its partner's by a half unit. This means that every bosonic particle must have a fermionic counterpart as a superpartner. The superpartner of an electron, for example, is a *selectron,* of a quark a *squark,* etc. As a bonus, *super-pairing* also averts the embarrassing appearance of vibrational patterns whose mass squared is bizarrely negative (tachyons).

12. For the sake of "*average energy conservation,*" these messenger particles are conceived as "virtual" particles in the sense that they emerge spontaneously and "live" for a split moment on "borrowed but returnable" energy. The

production of such particles is allowed by the uncertainty principle. Energy and time are conjugate variables and therefore subject to the reciprocal relation of quantum uncertainty. The energy in a chunk of field will therefore fluctuate more and more widely (increased uncertainty) as the space-time scale on which the region is examined gets smaller and smaller (i.e., becomes more and more specified). By virtue of mass-energy equivalence, this makes possible the emergence of a messenger particle (boson) and its antiparticle. Immediately later, these two annihilate each other so as to restore the borrowed energy. In this manner virtual electron-positron pairs can also be created, which, an instant later, annihilate each other and give up a photon that then vanishes as repayment of the borrowed energy.

13. For a more technical introduction to this vast subject, by one of its leading theorists, Edward Witten et al., see, M. B. Green, J. H. Schwarz, and E. Witten, *Superstring Theory* (Cambridge: Cambridge University Press, 1987).

Volume 1 starts out with quantum field theory and the motion of a classical point-particle moving in Minkowski space and then goes on to open and closed bosonic strings (i.e., strings having vibrational patterns with whole number spin). The account deals first with free strings, and later with interacting ones It then proceeds with various approaches to the further (supersymmetric) development of the subject. The discussion is intended to be self-contained but presupposes significant technical familiarity with quantum field theory. Substantial parts, however, are accessible to the more general reader.

Volume 2 deals with more advanced subjects and with phenomenological questions, i.e., issues of experimental confirmation. Particularly instructive, throughout, is the manner of approach, which, at various levels of exposition, starts with a classical description and then goes on to "quantumization" by introducing the appropriate *quantizing* conditions. "Quantumization" is a term of my own coinage and is not to be confused with the more specific term *quantization.*

14. The idea was at least partially inspired by the beta function, which correctly describes interactions at the nuclear level if we assume that the interacting particles are construed as vibrating strings. The beta function is a familiar mathematical formulation, already very useful for solving a wide variety of physical problems. Historically famous for its role in the development and generalization of the concept of factorial, it is very useful for evaluating a wide variety of definite integrals and, particularly, for establishing integral representations of a famous class of other functions, namely, Bessel functions, which, in turn, are useful in solving physical problems.

15. For example, the more the energy of vibration the more the mass.

16. In theoretical terms, whatever small mass might be associated with the miniscule extension of simple (i.e., unwrapped) strings, would be canceled by the quantum effects that occur at sub-Planck levels.

17. The various force charges of a particle are those properties of the particle that determine how it will be affected, respectively, by the fundamental forces of nature.

18. Recall that attempts to incorporate gravitation into quantum field theory resulted in unacceptable singularities, i.e., infinite probabilities, despite the mathematically "smoothing" effect of the superpairing required by supersymmetry. Well, the incorporation of gravity into string theory generates comparable absurdities. Unless strings are allowed nine degrees of vibrational freedom, i.e., nine spatial dimensions, some calculations will yield negative probabilities. At least ten dimensions, therefore—nine spatial and one time—are logically required. Adding to this complication is the fact that because we cannot in any way experience the extra dimensions they must be postulated as "curled up" at the level of sub-microdistances.

19. For a strongly unfavorable comment on string theory, along this and other lines of argument, see, Peter Woit, "Is String Theory Even Wrong?," *American Scientist*.90 (March 2002): 110–12. Another critic of string theory has been Nobel laureate Sheldon Glashow. For a good account of his views, see the interview of Glashow on the *Nova* Television Series (July 2003). A strong proponent of string theory and one of its leading architects is Edward Witten. For a very readable and favorable account see: "The Universe on a String," *Astronomy Magazine* (December 19, 2005).

20. Unfortunately, the mathematical difficulties of describing a vibrational string dynamics in the resulting ultra-microspace (known as a Calabi-Yau space) are severe enough to be beyond anyone's present-day abilities. As yet, therefore, no complete and well-formulated set of equations defines the theory. Nor is it assured, even were such a set of equations available, that useful solutions could be found—solutions that would indicate in which one of the infinitely many possible Calabi-Yau spaces the strings of string theory vibrate. Finally, the number of solutions could be enormously large with no clue as to which one (or ones) would uniquely describe our universe and square, for example, with the known facts of cosmology.

21. In this regard, one might note that in the four-dimensional physical theory of the familiar phenomenal world, the basic equations are typically second-order partial differential equations whose *exact* solutions are often not easily found and become virtually inaccessible when more than two interacting entities are involved. If this is so, imagine the complexity of dealing with physical contexts in ten spatial dimensions!

22. It is entirely possible that string theory, as presently construed, may be fundamentally too general. Indeed, it may turn out that without additional theoretical specifications, even a complete set of finished equations will yield an infinite number of exact solutions each describing a different possible universe. In such a case, singling out the solution that uniquely fits our universe would not be possible without further theoretical refinement.

23. No less exotic in the standard model is the behavior of the strong force that accounts for the bundling of quarks into familiar nuclear particles such as protons and neutrons but which, unlike the other fundamental forces, does not diminish with distance. While experiments with intense particle beams seem to be edging toward further enlightenment in this area, the strong force remains one of the hard challenges in particle physics. String theory which

aims to explain the manner of variation of all four fundamental forces is a sweeping theoretical promise in this regard.

24. Introducing the purely formal, quantum mechanical notion of spin into the formalisms of contemporary supersymmetric theory requires—as a matter of mathematical necessity—that every elementary particle and antiparticle of the standard model, whether of the bosonic or fermionic type, be paired with a corresponding supersymmetric partner with spin diminished by one-half.

For several years now, particle physicists at Fermilab have been trying to detect the superpartners of gluons (gluinos) whose existence is crucial for supersymmetry theory. These particles, however, have eluded the experimenters indicating that their masses may be too large to be created by the amount of energy the tevatron can deliver. The hunt, therefore, has gone to the Large Hadron Collider at CERN.

25. It may appear here that, in being ontologically expansive, the uncertainty principle has explanatory force. On the contrary, the principle, as a theorem of QT, tells us only that for ultra-small distances, the energy fluctuations are large enough to make the production of particle antiparticle pairs *possible* (mass-energy equivalence). QT, therefore, in remaining algorithmic throughout, does not specify causes. It merely *allows* the probable production of such particles.

26. QT as an algorithmic nomological offers no basic explanatory ontology. Spin, therefore, is properly viewed not as something that QT positively attributes to elementary particles but rather as a property that QT *allows* for both for the point-particles of QFT and the strings of string theory.

27. After E. Bogomoln'yi, M. Prasad, and C. Summerfield—the three who discovered them.

28. The calculable states are states with the smallest possible mass for a given amount of electric charge.

29. Again, symmetry (of a kind) takes center stage—this time, however, not imposed from without, but begotten from an already operative symmetry, i.e., supersymmetry.

30. The probability of loop formation due to the production of virtual string pairs increases as the coupling constant increases. For coupling constants greater than one the perturbing influence on the original interaction tends to invalidate the perturbative approach, which presupposes no more than a *few* (perturbing) splits and "rejoinings" subsequent to the initial interaction.

31. In the context of today's technology one would need accelerators of cosmologic proportions in order to bring the colliding particles to the energy levels needed for detecting strings. The technical impossibility of meeting this requirement, therefore, seems insuperable unless other methods of concentrating (focusing) energy may become available for achieving the desired results with terrestrial technologies.

32. A less formal, operational account is sometimes given for reconciling a quantumized string theory with relativity. It goes something like the following: Since strings are the elements of all physical reality, any probing mechanism

must ultimately consist of strings. But strings cannot probe structures smaller than strings, i.e., sub-Planck structures, where all the quantum turbulence and discontinuities are. The quantum mishmash, therefore, is not observable and therefore has no physical significance. Operationally speaking, therefore, the conflict between QT and relativity is seen not to arise in string theory.

33. This is not to say that string theory, at its present level of development, has no specific observational consequences at all. The theory, for example, allows *fractions* of what we classically take to be the elemental electronic charge; so that any discovery of these would constitute a dramatic, though limited piece of experimental confirmation. And, as we have already noted, there are still other consequences of the theory that may be experimentally accessible such as the existence of selectrons, squarks, or any of the superpartners.

34. Specialists report that, as may be calculated, a change of only half a percent in the strong force (operative in nuclear structure) would be enough to prevent the formation of carbon and oxygen needed for the production of biological material. And, only a slight variation in the fine structure constant on which the strength of the electromagnetic force depends would compromise the formation of molecules.

Chapter 9. Beyond Quantumization

1. It is the linearity of the Schrödinger formalism that entails superposition, a formal property deeply connected with the measurement problem. Departing from linearity could, therefore, be seen as a major move toward solving the measurement problem.

2. Present-day astronomical data strongly suggest that (vastly) most of the matter/energy is of a dark or invisible and exotic form constituting about 96 percent of the total in the universe. Quantum theoretic calculations, however, come to wildly different results. They yield, for example, a value of the order of 1060 for the ratio of dark to visible matter.

3. Some theorists see no possibility of resolving the measurement problem without considerable supplementation and modification of the Schrödinger formalism. See, for example, Zeh (2003).

4. See: D. Bohm and J. Bub, "A Proposed Solution to the Measurement Problem in Quantum Mechanics by a Hidden Variable Theory," *Reviews of Modern Physics* 38 (1966): 453. For some brief accounts, see: J. Polchinsky, "Weinberg's Nonlinear Quantum Mechanics and the Einstein-Podolsky-Rosen Paradox," *Physical Review Letters* 66 (Jan. 1991): 397–400. Also see: L. L. Bonilla and F. Guinea, "Reduction of the Wave Packet Through Classical Variables," *Physics Letters* B271 (1991): 196–200.

5. E. P. Wigner, "Remarks on the Mind-Body Question," in *The Scientist Speculates*, ed. I. J. Good (New York: Heinemann and Basic Books, 1962).

Bibliography

Albert, D., "Special Relativity as an Open Question." In *Relativistic Quantum Measurement and Decoherence,* ed. H. P. Breuer and F. Petruccione. Heidelberg: Springer, 1999.

Albert, D., and B. Loewer. "Interpreting the Many Worlds Interpretation." *Synthese* 77 (1988): 195–213.

Aspect, A., et al., *Physical Review Letters* 47 (1981): 460; *Physical Review Letters* 49, 91 (1982): 1804

Bacciagaluppi, G. "The Role of Decoherence in Quantum Theory." *Stamford Encyclopedia of Philosophy* (Winter 2003). Edited by E. N. Zalta. This is an introductory unspecialized account.

Barret, J. *The Quantum Mechanics of Minds and Worlds.* Oxford: Oxford University Press, 1999.

Bell, J. S., *Physics* 1 (1964): 195.

———. *Speakable and Unspeakable in Quantum Mechanics.* Cambridge: Cambridge University Press, 1987.

Bohm, D. *Causality and Chance in Modern Physics.* London: Routledge and Kegan Paul, 1957.

———. "A Suggested Interpretation of Quantum Theory in Terms of Hidden Variables." *Physical Review* 85 (1952): 166–79.

———, with J. Bub. "A Proposed Solution to the Measurement Problem in Quantum Mechanics by a Hidden Variable Theory." *Reviews of Modern Physics* 38 (1966): 453.

Bohr, N. "Discussion with Einstein on Epistemological Problems in Atomic Physics." In *Albert Einstein, Philosopher Scientist,* ed. Paul A. Schilpp. New York: Harper and Row, 1949.

Bonilla, L. L., and F. Guinea. "Reduction of the Wave Packet through Classical Variables." *Physics Letters* B271 (1991): 196–200.

Bub, J. "Hidden Variables and the Copenhagen Interpretation—A Reconciliation." *British Journal for the Philosophy of Science* 19 (1968).

Cannavo, S. *Nomic Inference (An Introduction to the Logic of Scientific Inquiry).* The Hague: Martinus Nijhoff, 1974.

———. *Think to Win.* Amherst, NY: Prometheus Books, 1998.

Chalmers, D. J. *The Conscious Mind.* Oxford: Oxford University Press, 1996.

d'Espagnat, B. "The Quantum Theory and Reality." *Scientific American* 241, no. 5 (1979).

DeWitt, B. S. "Quantum Mechanics and Reality." *Physics Today* (1970): 30–35.

———, and N. Graham. *The Many Worlds Interpretation of Quantum Mechanics.* Princeton: Princeton University Press, 1973.

Dirac, P. *The Principles of Quantum Mechanics.* Oxford: Clarendon Press, 1958.

Einstein, A., B. Podolsky, and N. Rosen. "Can Quantum Mechanical Description of Physical Reality Be Considered Complete?" *Physical Review* 47 (1935).

Everett, H., "Relative State Formulation of Quantum Mechanics." *Reviews of Modern Physics* 29 (1957): 454–62.

Ghirardi, G. C. "The Quantum Picture of Natural Phenomena." Preprint; Trieste: International Centre for Theoretical Physics, 1993.

———. "Collapse Theories." *Stanford Encyclopedia of Philosophy* (January 28, 2007).

———, A. Rimini, and T. Weber. "A Model for a Unified Quantum Description of Macroscopic and Microscopic Systems." In *Quantum Probability and Applications,* ed. L. Acardi, et al. Berlin: Springer Verlag, 1985.

———. "Unified Dynamics for Microscopic and Macroscopic Systems." *Physical Review* D34 (1986): 470.

Giulini, D., et al., eds., *Decoherence and the Appearance of a Classical World in Quantum Theory.* Heidelberg: Springer, 1996, ch. 2, 16.

———. "Decoherence, Einselection, and the Existential Interpretation." *Philosophical Transactions of the Royal Society of London* A 56 (1998): 1793–1820.

Glashow, S. Interview. *Nova.* July 2003.

Goldstein, H. *Classical Mechanics.* Reading, MA: Addison-Wesley, 1959.

Greene, B. *The Elegant Universe.* New York: W.W. Norton, 1999.

Heisenberg, H. *The Principles of Quantum Mechanics.* Oxford: Oxford University Press, 1958.

Hempel, C. G., and P. Oppenheim. "Studies in the Logic of Explanation." *Philosophy of Science* 15 (1948): 135–78.

Hemmo, M., and I. Pitowsky. "Probability and Nonlocality in Many Minds Interpretations of Quantum Mechanics." *British Journal for the Philosophy of Science* 54 (2003): 225–44.

Herbert, N. *Quantum Reality.* New York: Anchor Books, 1985.

Hogan, J. *The End of Science.* New York: Broadway Books, 1997.

Jammer, M. *The Conceptual Development of Quantum Mechanics.* New York: McGraw-Hill, 1966.

———. *The Philosophy of Quantum Mechanics.* New York: John Wiley and Sons, 1974.

Jauch, J. M., and C. Piron. "On the Structure of Quantal Proposition Systems." *Helvetica Physica* Acta 42 (1969): 842–48.

———. "Can Hidden Variables Be Excluded from Quantum Mechanics?" *Helvetica Physica* Acta 36 (1963): 827–37.

Joos, E., et al. *Decoherence and the Appearance of a Classical World in Quantum Theory.* Heidelberg: Springer, 2003.

Kitcher, P. S. "Philosophy of Science." *Britannica Online Encyclopedia,* www.britannica.com/eb/print?articleleId=108542 &fullArticle= true& tocId=271817.

Lederman, L. *The God Particle.* New York: Dell, 1993.

Leggett, A. J. "Schrödinger's Cat and Her Laboratory Cousins." *Contemporary Physics* 25 (1986): 583–94.

Lewis, D. "How Many Lives Has Schrödinger's Cat?" *Australasian Journal of Philosophy* 82, no. 1 (2004): 3–22.

Lockwood, M. "'Many Minds Interpretation of Quantum Mechanics." *British Journal for the Philosophy of Science* 47 (June 1996): 159–88.

Loewer, B., "Comment on Lockwood." *British Journal for the Philosophy of Science* 47 (1996): 229–332.

Myrvold, W. C. "Relativistic Quantum Becoming." *British Journal for the Philosophy of Science* 54 (2003): 475–507.

Pagels, H. R. *The Cosmic Code.* New York: Simon and Schuster, 1982.

Papineau, D., "Many Minds Are No Worse than One." *British Journal for the Philosophy of Science* 47 (1996): 234–41.

Pearle, P. "Reduction of a State Vector by a Nonlinear Schrödinger Equation." *Physical Review* D13 (1979).

Penrose, Roger. *Shadows of the Mind.* Oxford: Oxford University Press, 1994.

Piron, C. *Helvetica Physica* Acta 36 (1963).

Polchinsky, J. "Weinberg's Nonlinear Quantum Mechanics and the Einstein-Podolsky-Rosen Paradox." *Physical Review Letters* 66 (Jan. 1991): 397–400.

Putnam, H., "A Philosopher Looks at Quantum Mechanics." In *Mathematics, Matter, and Method. Philosophical Papers,* Vol. 1. Cambridge: Cambridge University Press, 1979.

Reichenbach, H. "Foundations of Quantum Mechanics." In *Hans Reichenbach: Selected Writings,* ed. R. S. Cohen and M. Reichenbach. Dordrecht and Boston: Reidel, (1978), 253–54.

Salmon, W. *The Foundations of Scientific Inference.* Pittsburgh: University of Pittsburgh Press, 1964.

———. *Scientific Explanation and the Causal Structure of the World,* Princeton: Princeton University Press, 1984.

Schrödinger, E. "Discussions of Probability Relations between Spacially Separated Systems." *Proceedings of the Cambridge Philosophical Society* 31 (1935): 555-63.

Slater, J. *Quantum Theory of Matter.* New York: McGraw-Hill, 1951.

Stapp, H. P. *Mind Matter and Quantum Mechanics.* Heidelberg: Springer, 1993.

Tittel, W., et al. "Violation of Bell Inequalities by Photons More than 10 km Apart." *Physical Review Letters* 81 (1998): 3563–66.

Wallace, D. "Worlds in the Everett Interpretation." *Studies in the History and Philosophy of Modern Physics* 33, no. 1 (2002): 637–61.

———. "Everett and Structure." *Studies in the History and Philosophy of Modern Physics* 34, no. 1 (2003): 87–105.

Wheeler, J. A. "Assessment of Everett's Relative State Formulation of Quantum Theory." *Reviews of Modern Physics* 29 (1957): 463–65.

Wheeler, J., and H. Zureck., eds. *Quantum Theory and Measurement.* Princeton: Princeton University Press, 1987.

Wick, G. C., A. S. Wightman, and E. P. Wigner. "The Intrinsic Parity of Elementary Particles." *Physical Review* 88, no. 36 (1952): 101–105.

Wightman, A. S. "Superselection Rules; Old and New." *Il Nuovo Cimento* 110B (1995):751–69.

Wigner, E. P. "Remarks On the Mind-Body Question." In *The Scientist Speculates,* ed. I. J. Good. New York: Heinemann, 1962, 284–301.

Witten, E. "The Universe on a String." *Astronomy Magazine* (December 19, 2005).

———, M. Green, and J. H. Schwarz. *Superstring Theory.* Cambridge: Cambridge University Press, 1987.

Woit, P. "Is String Theory Even Wrong?" *American Scientist* 90 (March 2002): 110–12.

Yang, C. N., and R. L. Mills. "Conservation of Isotopic Spin and Isotopic Gauge Invariance." *Physical Review* 96 (1954): 191–95.

Zeh, H. D. "The Problem of Conscious Observation in Quantum Mechanical Description." *Epistemological Letters* 63 (1981).

Zurek, H. "Basic Concepts and Their Interpretation." In *Decoherence and the Appearance of a Classical World in Quantum Theory,* ed. D. Giulini et al. Heidelberg: Springer Verlag, 1996.

———. "Decoherence, Einselection, and the Existential Interpretation." *Philosophical Transactions of the Royal Society of London* A 356 (1998).

———."Decoherence, Einselection, and the Quantum Origins of the Classical." *Review of Modern Physics* 75 (July 2003).

Index